Resonances of *El Chavo del Ocho* in Latin American Childhood, Schooling, and Societies

New Directions in Comparative and International Education
Edited by Stephen Carney, Irving Epstein and Daniel Friedrich

This series aims to extend the traditional discourse within the field of Comparative and International Education by providing a forum for creative experimentation and exploration of alternative perspectives. As such, the series welcomes scholarly work focusing on themes that have been under-researched and under-theorized in the field but whose importance is easily discernible. It supports works which theoretical grounding is centered in knowledge traditions that come from the Global South, encouraging those who work from intellectual horizons alternative to the dominant discourse. The series takes an innovative approach to challenging the dominant traditions and orientations of the field, encouraging interdisciplinarity, methodological experimentation, and engagement with relevant leading theorists.

Forthcoming title in the New Directions in Comparative and International Education series

Education in Radical Uncertainty, Stephen Carney and Ulla Ambrosius Madsen

Also available from Bloomsbury

Education in Mexico, Central America and the Latin Caribbean, edited by C. M. Posner, Christopher Martin and Ana Patricia Elvir
Education in South America, edited by Simon Schwartzman

Resonances of *El Chavo del Ocho* in Latin American Childhood, Schooling, and Societies

Edited by Daniel Friedrich and Erica Colmenares

Bloomsbury Academic
An imprint of Bloomsbury Publishing Plc

B L O O M S B U R Y
LONDON • OXFORD • NEW YORK • NEW DELHI • SYDNEY

Bloomsbury Academic
An imprint of Bloomsbury Publishing Plc

50 Bedford Square
London
WC1B 3DP
UK

1385 Broadway
New York
NY 10018
USA

www.bloomsbury.com

BLOOMSBURY and the Diana logo are trademarks of Bloomsbury Publishing Plc

First published 2017

British Library Cataloguing-in-Publication Data
A catalogue record for this book is available from the British Library.

ISBN: HB: 978-1-4742-9890-2
ePDF: 978-1-4742-9887-2
ePub: 978-1-4742-9888-9

Library of Congress Cataloging-in-Publication Data
A catalog record for this book is available from the Library of Congress.

Cover image © Long Quattro/iStock

Typeset by RefineCatch Limited, Bungay, Suffolk

Contents

Notes on Contributors vii
Series Editors Foreword x

Introduction: Resonances of *El Chavo del 8* in Latin American Childhood, Schooling, and Societies *Daniel Friedrich and Erica Colmenares* 1

Plateau I El Chavo Encounters Latin American Education and Childhoods

1 Schooling, Pedagogical Imaginaries, and Latin American Childhoods in *El Chavo del 8* *Nicolás Arata and Daniel Friedrich* 19

2 Conceptions of Childhood in the *Vecindad* *Victoria Parra-Moreno* 35

3 "Here comes Chavo! Everyone's watching the TV"... Thinking About Difference and Alterity, Childhood, and Education *Ana Paula Marques de Carvalho and Rita de Cássia Prazeres Frangella* 51

Plateau II El Chavo's Encounters with Other Latin American Societies and Cultural Artifacts

4 From the Picaresque Novel to *El Chavo del 8* *Carlos Aguasaco* 69

5 Border-crossing Chespirito: El Chavo del 8 Meets Pepito in Exile *Limarys Caraballo* 87

6 *El Chavo del 8* as an "Intimate Public" in Venezuela: What Happened to the Good Life? *Erica Colmenares* 111

Plateau III El Chavo's *Media*-ted Encounters

7 Figures of Mexican State, Society, and Subject in Chespirito's TV *Ernesto Treviño Ronzón* 131

8 Media Education "*Sin Querer Queriendo*" *Dulce María Cabrera and José Carbajal Romero* 155

Coda *El Chavo del 8's* Connections and Reverberations
Erica Colmenares and Daniel Friedrich 177

Glossary 187
Index 189

Notes on Contributors

Carlos Aguasaco is Associate Professor of Latin American Cultural Studies and Spanish at The City College of New York (CUNY), USA. As author and scholar, he has been invited to present his work in universities and cultural institutions of the US, Mexico, Cuba, Colombia, Dominican Republic, Puerto Rico, Honduras, Ecuador, Brazil, and Spain. He is the author of *¡No Contaban con mi astucia! México: parodia, nación y sujeto en la serie de El Chapulín Colorado* (2014). His main academic interests are: literary theory, media studies, contemporary poetry, Latin American popular culture, and residual ideologies of the Spanish Golden Age.

Nicolás Arata coordinates the graduate program at Consejo Latinoamericano de Ciencias Sociales (CLACSO), Argentina. He teaches the History of Argentine and Latin American Education at the Universidad de Buenos Aires, Argentina, and Universidad Pedagógica and has previously researched and taught in Spain, Germany, and most of Latin America. He holds doctorates from the Universidad de Buenos Aires, Argentina, and DIE-CINVESTAV, Mexico. He has published books, chapters, and articles in both domestic and international journals and is the academic director of the *Argentine Society for the History of Education Yearbook*.

Dulce María Cabrera is a professor at the Benemérita Universidad Autónoma de Puebla, Mexico. She holds a PhD in Pedagogy from the Universidad Nacional Autónoma de México, Mexico. Her areas of interest are the political analysis of discourse and education, the relationship between academic experiences and labor, and policy, curriculum, and educational management.

Limarys Caraballo is Assistant Professor of English Education at Queens College-CUNY, USA, and Research Fellow at the Teachers College Institute for Urban and Minority Education, USA. Her scholarship promotes educational justice via inquiries into students' cultural identities and literacies, curriculum theory, and youth. Her research has been recognized by AERA, the American Association for Teaching and Curriculum, and the National Council for Teachers of English. Her work has been published in journals including the *Journal of Curriculum and Pedagogy*, *Urban Education*, and *Review of Research in Education*;

she is co-author of *Policy, Professionalization, Privatization, and Performance Assessment: Affordances and Constraints for Teacher Education* (2016).

José Carbajal Romero holds a Doctorate and Masters Degree in Science from the Centro de Investigación y de Estudios Avanzados (Cinvestav) at the Instituto Politécnico Nacional, Mexico. He also holds a degree in Communication Science from the Universidad Autónoma Metropolitana, Mexico. He has published many studies about the educational uses of Information and Communication Technologies (ICT) and has researched topics related to contemporary technological and inquiry dimensions in the ICT's role producing subjectivity.

Erica Colmenares is a doctoral candidate in the Curriculum and Teaching department at Teachers College, Columbia University, USA. As an instructor and field supervisor at Teachers College Elementary Inclusive Preservice program, her research interests include preservice teacher education, learning to teach for social justice, and theories of affect. Her current research involves exploring the affective realms of social justice-oriented teacher education programs. She has published in *Education Policy Analysis Archives* and *Teachers College Record.*

Rita de Cássia Prazeres Frangella is Professor of the Graduate Program in Education in the School of Education at Rio de Janeiro State University (UERJ). She is a Young Scientist of Our State (FAPERJ). Her research focuses on curriculum policies as cultural productions from a discursive perspective. She coordinates research on curriculum policies of teacher training, early childhood education and early years of basic education. She is the Secretary General of the Brazilian Association of Curriculum (ABdC), and a member of the Scientific Committee of the National Association of Graduate Studies and Research in Education (ANPEd).

Daniel Friedrich is Associate Professor of Curriculum at Teachers College, Columbia University, USA. Prof. Friedrich has published extensively in the fields of Curriculum Studies, Teacher Education, and Comparative and International Education. His book *Democractic Education as a Curricular Problem* was published in 2014. He is the co-founder of the Post-Foundational Approaches to Comparative and International Education Special Interest Group at the Comparative and International Education Society, and one of the co-editors of the series *New Directions in Comparative and International Education* for Bloomsbury Academic.

Victoria Parra-Moreno is a doctoral candidate in Curriculum and Teaching concentrating in early childhood policy at Teachers College, Columbia University, USA. She holds an M.A. in Psychology from the Universidad de Chile, Chile, as well as an Ed.M. in Curriculum and Teaching from Teachers College, Columbia University, USA. She currently works as an independent consultant/researcher in education and early childhood. She has worked previously in the areas of education policy design, implementation, and evaluation. Her research interests focus on early childhood education, public policies, and public education.

Ana Paula Marques de Carvalho is a doctoral candidate in Education at the Rio de Janeiro State University (UERJ), Brazil. She holds a Masters in Education and coordinates courses, events, and publications at the Extension Department at Rio de Janeiro State University, Brazil.

Ernesto Treviño Ronzón is a full-time researcher at the University of Veracruz, Mexico. He is coordinator of the research group "Sociopolitical Studies" and dictates seminars in areas such as research methodology, public policies, and discourse analysis. His interests include the intersection of policy, politics, and policy in fields such as education and knowledge following post-structural perspectives. He is also interested in the study of violence. He has dictated conferences and has been a visiting scholar in universities in Mexico, the USA, Argentina, and others. He participates in scientific organizations of different scopes such as COMIE (Mexico), LASA and AERA (USA), CLACSO (Latin America).

Series Editors Foreword

The field of comparative and international education requires its researchers, teachers, and students to examine educational issues, policies, and practices in ways that extend beyond the immediate contexts with which they are most accustomed. To do so means that one must constantly embrace engagement with the unfamiliar, a task that can be daunting because authority within academic disciplines and fields of study is often constructed according to convention at the expense of imagination and creativity. Comparative and international education as an academic field is rich and eclectic, with a long tradition of theoretical and methodological diversity as well as an openness to innovation and experimentation. However, as it is not immune to the conformist—especially disciplinary—pressures that give academic scholarship much of its legitimacy, we believe it important to highlight the importance of research and writing that is creative, thought-provoking and, where necessary, transgressive. This series offers comparative and international educators and scholars the space to extend the boundaries of the field, encouraging them to investigate the ways in which under-appreciated social thought and theorists may be applied to comparative work and educational concerns in new and exciting ways. It especially welcomes scholarly work that focuses upon themes that have been under-researched and under-theorized but whose importance is easily discernible. It further supports work whose theoretical grounding is centered in knowledge traditions that come from the Global South and welcomes perspectives including those that are associated with post-foundational theorizing, non-Western epistemologies, and performative approaches to working with educational problems and challenges. In these ways, this series provides a space for alternative thinking about the role of comparative research in reimagining the social.

It is thus fitting that *Resonances of* El Chavo del Ocho *in Latin American Childhood, Schooling, and Societies* is the initial volume in the *New Directions in Comparative and International Education* series. Its basic concept is that the enduring popularity of a television program broadcast throughout Latin America for many decades can best be understood by examining the plasticity with which its images are reinterpreted within multiple political, social, and cultural contexts. Through relying upon theoretical frameworks that draw from literary, cultural, media studies, and social theory, *Resonances of* El Chavo del Ocho raises important questions regarding the relationship between childhood and modernity, the nature of social class and its impact upon dependency and futuristic aspiration, and the connections between family, schooling, and community. It thus casts new light upon conventional analyses of the possibilities for agency in the midst of neoliberalism and does so by drawing upon an eclectic sampling of theory and method. While comparisons can inevitably be drawn with conventional television programs and other media forms emanating from the Global North, this is very much a project with a Global South focus. One of the most exciting attributes about the volume is its honesty in showcasing perspectives that are at variance with one another and are occasionally contradictory. It remains for the reader to sort out the very real ambiguities that color our interpretations of text and image.

Irving Epstein and Stephen Carney

Introduction

Resonances of *El Chavo del 8* in Latin American Childhood, Schooling, and Societies

Daniel Friedrich and Erica Colmenares

¡Fue sin querer queriendo! ¡Se me chispoteó! Eso, eso, eso. ¡Es que no me tienen paciencia! Mention any of these catchphrases to almost any Latin American aged five and up and brace yourself for an impassioned account of the person's experiences with and alongside *El Chavo del 8*, possibly the most famous creation of Mexican author/comedian/director/producer/actor Roberto Gómez Bolaños (aka Chespirito, a play on the Spanish pronunciation of Shakespeare). The television show, which at the peak of its popularity in the mid-1970s reached an approximate audience of 350 million per episode across the Americas, has left an indelible mark on the lives of multiple generations across the region, and is still extremely popular through reruns, spinoffs, and merchandise.

El Chavo del 8 is a character unlike any other, although singling out the reasons for its success seems like an impossible task. The sketch's setup consists of a small group of adult actors, most of them acting as children, who live in a Mexican *vecindad*, a lower-class housing complex. El Chavo himself is an orphan who lives in a wine barrel (although he states now and then that he actually lives in Apartment 8, a fact that is never confirmed in the show). The character most people fear or despise is the landlord. The humor relies on a mix of physical comedy and verbal jokes, with a simplicity that appeals to the youngest audiences, yet engages older viewers through nostalgia and tenderness. This seemingly uncomplicated format, however, has produced an unprecedented phenomenon, warranting creative approaches to account for its effects on producing particular visions of Latin American childhood, schooling, and societies.

This edited volume challenged different scholars from Latin America to engage in thinking about images of schooling and childhood, Latin American-ness, affect, media, class, and other such topics through *El Chavo's* narratives, images, sounds, and experiences of viewership. There are not many instances where the same cultural product has traveled through Latin America, leaving a lasting impact of several decades, yet with specific effects because of the unique ways it was received in each location, at different times, in different political moments. The book invited reflection from an international perspective, without attempting to compare or reach consensus on any ultimate meanings of the work. Instead, it was conceived more as a rhizomatic approach to reading *El Chavo*; *El Chavo* might be this, *and* that, *and* that, *and* that ... Exploring the connections between visual culture studies and transcultural curriculum studies, the chapters contained in this volume navigate a space of transnational poetics of living with popular culture.

This introduction will provide readers unfamiliar with the show with a brief description of the main characters, settings, and logics, as well as a discussion of the context of emergence of the show and its spread through the continent, as a way of partly framing the chapters. Given the focus of the book, Roberto Gómez Bolaños's biography will be barely touched upon, privileging instead a theoretical framework that will allow us to see *El Chavo* and its effects in their multiplicity, their tensions and contradictions, their entanglements with and against power, acknowledging always the impossibility of taking a fully analytic approach with regard to such an affectively loaded product.

Both of us editors are Latin Americans, part of a generation that grew up in the 1980s and thus deeply implicated in the production of subjectivities tied to *El Chavo*. We participated in the Latin America that allowed *El Chavo* to emerge as a continental phenomenon, while we are also aspects of the Latin America produced by *El Chavo*. This applies to all the contributors and, as such, this introduction—and the book as a whole—could never play into the modernist myth of an object of study that is distinct from the observer, an objective analysis in which the researcher is forever struggling to distance *him*self from both the analysis and *his* own persona. While this myth has been contested for over half a century, the study of pop culture adds a new layer to this conversation. Unless one is willing to situate oneself outside the people that constitute the "popular" in pop culture—something neither one of us is willing to do—one is necessarily implicated in the phenomena one is studying. Analyzing a show like *El Chavo del 8*, a show that was not influential somewhere else, for other people, but that we ran back from school to watch, that was an integral part of our childhoods, is not an obstacle, but a challenge to deploy alternative theoretical lenses. These frameworks need to account for our own experiences, the ways in which catchphrases and body ticks move(d) us, the pleasure we gained from viewing the same gags over and over, the tensions we embodied in laughing with and about an extremely poor child who lived in a barrel, among others. In this sense, our interest somewhat mirrors that of Elizabeth Chin (2016), who in her own consumer diaries is fascinated by Marx's collection of napkins and his affective connections to these commodities while still critiquing the capitalist order that produced them in the first place.

What was *El Chavo del 8*?

El Chavo del 8 was created by Roberto Gómez Bolaños in 1971, as one of a set of characters that he would portray in his show *Chespirito*. All of Gómez Bolaños's characters tended to be noble, yet demonstrated failed or flawed traits. El Chapulín Colorado, for example, was a parody of American superheroes: "More agile than a turtle . . . more noble than lettuce . . . his logo is a heart!," announced a deep voice as Chespirito's tiny frame jumped around in red tights, ready to embark on ridiculous adventures where he would more often than not end up being rescued by the "victims" that had called on him for help. El Chómpiras was a petty thief who never managed to get away with any bounty, El Doctor Chapatín made his patients sicker than the lack of treatment they were enduring until he arrived, and Chaparrón Bonaparte refused to go back to the mental institution he had escaped from. Yet no other character hit audiences as hard or as deep as El Chavo del 8. El Chavo was perhaps the least outlandish of Chespirito's creations: an extremely poor orphan, always with good intentions and an empty stomach as his main motivator. No powers, mental health issues, or criminal records. Just a poor kid from the *vecindad*.

El Chavo was also the only sketch of Chespirito's in which all the other characters became independently recognizable household names. La Chilindrina, the wise-cracking girl in love with him, tricked everybody to get what she wanted. Her single father, Don Ramón, never held a stable job and strived to escape from the landlord, always lacking money to pay the rent. Quico lived with his mom, Doña Florinda, and struggled to retain the remnants of a former middle-class life with new toys that were paraded in front of other kids, perhaps to cover for the fact that they were living in the same *vecindad* as *la chusma* [the rabble]. Profesor Jirafales, the schoolteacher, was the only character who effectively possessed the cultural capital that distinguished him from the inhabitants of the *vecindad*, yet his prudish love story with Doña Florinda always kept him close. The landlord, Señor Barriga, symbolized wealth through a name that reflected his rotund body shape; his son, Ñoño, brought his higher economic capital to bear in every interaction with the rest of the children. Finally, we have Doña Clotilde, or La Bruja del 71, who—while never given much of a back story—was an older woman and the subject of numerous ageist jokes.

This ensemble inhabited *la vecindad*, which became the other living, breathing character in the show. *Vecindades* were housing complexes in which a central patio was surrounded by slum apartments. They had shared bathrooms and laundry rooms, and tended to be populated by renters who could barely make ends meet (for a fuller description of the history and role of *vecindades*, see Cabrera and Carbajal Romero in this volume). The show rarely left this setting, with the exception of skits that took place in El Chavo's school and an unusual pair of episodes in Acapulco. After watching hundreds of episodes that took place in the housing complex, we all became part of the *vecindad*—it became another place for us to hang out.

El Chavo del 8, as a character, is surrounded by mysteries. Two of them, linked to his moniker, stand out. First, Chavo is not a name, but a term for "lad." No one knows his name, and on the few occasions where he is about to say it, he is interrupted and both the other characters and the audience are left with an unsolved puzzle. Second, where did he live? Every time El Chavo got sad or lonely, he retreated into a wine barrel. On many occasions he is seen yawning and stretching as he comes out of the barrel, having just woken up from a nap. However, he insists every now and then that the barrel is not his home, but rather apartment 8.[1] Who does he live with? Who pays the rent? How did he get there? Is this all part of his imagination, or a way of avoiding the shame of admitting he is homeless? Yet a mystery of a different kind encouraged us to edit this volume: How is it possible for a character with these features to have moved, affected, and mostly entertained (sometimes reducing us to tears of laughter) swaths of generations across time and space?

The emergence of El Chavo as a phenomenon requires some contextualization and a set of caveats. While any attempt at setting up a scene for a whole continent across a decade could seem helpful for an unfamiliar readership, as incomplete as it may be, we want to resist the temptation to relate context to effect in a causal relation. We do not exactly know why El Chavo became such a success, and providing competing narratives about Latin America in the 1970s is not our way of answering that question, but merely a way of situating the show within a grid with its own sets of absences and presences (Puiggrós, 1996) based on our own contingent limitations and preferences.

Politically, Latin America was going through one of its most unsettling decades, which for this particular region, is saying quite a bit. Locked in a cycle of guerrilla struggles, bloody coups, authoritarian regimes, civil conflicts, changing economies, and foreign intervention, the continent was undergoing a deep change. Among the most significant was the mass migration of populations within and across borders. Within national borders, a rapid process of urbanization was taking place, with the continent's main cities growing exponentially as industrialization and modern farming technologies, coupled with specific policies, forced enormous numbers of rural farmers to move into the urban centers.[2] Across borders, significant portions of many countries' intellectual and political sectors were forced into exile by authoritarian regimes leading many of them to relocate in neighboring nations. We wonder, along with Berger, Berger, and Kellner (1974), whether in times of social upheaval the need for some kind of common experience, and/or escapism, could be a reason for the emergence of the El Chavo phenomenon. (For a personal account of cross-border migration in relation to El Chavo, see Caraballo in this volume.)

More specifically, the 1970s saw a resurgence of discourses of Latin American brotherhood as one way to counter Cold War binaries, a growing Americanization of global popular culture and consumption, threats to regional sovereignty, and a development discourse that often called for an acritical severing of the continent's ties to its past and tradition in order to become fully modernized. What better then, to unite La Patria Grande, than a common language and set of cultural references, a character that was familiar to all but universal enough to overlook specificity? Yet beyond the appeal of the seemingly common, there is another layer to this relation. According to Berger, Berger, and Kellner (1974), modernizing processes have left vast sectors of the population "'alienated' and beset with the threat of meaninglessness" (p. 195), *homeless* in regard to the perceived safety of former communal ties. Here comes a homeless boy—one that while extremely poor and constantly hungry mobilizes his fictional nature to avoid addressing the most disconcerting consequences of poverty and homelessness—who finds his home in each viewer's living room. El Chavo is, in some ways, the ultimate exile, kicked out of a grim reality to look for a new place to, at least temporarily, settle in, as a romanticized version of the intolerable abject.

The rise of Televisa (Televisión Via Satélite SA) as a media conglomerate cannot be understated as another contextual aspect. Televisa was on its way to becoming the largest multimedia company in the Spanish-speaking world; Chespirito's productions both benefitted from Televisa's resources and constituted one of the emerging empire's biggest assets. Technologically, the corporation outpaced its competitors in its early days: the implementation of a videotape system that replaced the more common (but costly) kinescope and 16 mm film, along with the acquisition of two video recording machines that made it possible to create multiple copies of any television show, facilitated the creation of a wide-reaching network. Such was its reach that by the early 1970s, Televisa had become the main exporter of television programs to the rest of Latin America (Aguasaco, 2014).

Politically, Televisa and *El Chavo* seemed to have a somewhat symbiotic relationship (see Treviño Ronzón, this volume). Mexico was experiencing one of its hardest cyclical political crises, with the Partido Revolucionario Institucional (PRI) continuing its hegemonic dominance over the country's politics it had initiated in 1929 and which would continue, uninterrupted, until 1989. However, the PRI was severely criticized for its role in the massacre of students in Plaza de las Tres Culturas de Tlatelolco in 1968. The government perceived a threat from a leftist uprising, and required a re-framing of the poor and the working class, a narrative—not new, but re-invigorated—of harmony and goodwill, of imagined simpler times when the needy would accept their social position as part of a natural order. *El Chavo* seemed, at least superficially, to fit the bill (although many of the chapters in this book will complicate this notion). We are not arguing here that the PRI had anything to do with Chespirito's show directly, particularly in terms of funding or writing scripts. Instead, our point is that there was a synergy that might help to explain the privileged slot the series had on Televisa's prime time (especially given the tight relationship between the media conglomerate and the PRI (Villamil, 2012)), and the ease with which the show fit into a certain discourse circulating in Mexico at the time. What this convergence cannot explain, however, is the persistence of El Chavo as a cultural icon across time and space, surviving left- and right-wing administrations, economically stronger and weaker periods, American pop culture dominance, regional or local resurgences of the show, and even technological advances that have thrown old-fashioned TV's relevance into question.

El Chavo as rhizome

There is probably not a single factor that can explain *El Chavo's* staying power, and we do not intend to even seek a complex explanation for this. Instead, we propose to understand Chespirito's show from a rhizomatic perspective. Our (and the rest of the authors in this volume's) analysis of *El Chavo* displaces the focus on the one "real" meaning of the object from which several consequences and effects can be traced, and instead seeks to multiply *El Chavo* in a non-hierarchical manner resisting all sorts of closures, inviting instead a continuous opening of new twisted paths and connections. Such thought, pioneered by Deleuze and Guattari (1987), implies a shift from positivist and post-positivist projects that, based on a modernist foundation, seek to get at the root of a correspondence between the word and the world, between language and reality. A rhizomatic encounter with *El Chavo*, seeks instead to multiply words and worlds and to disturb the hierarchical relationship between language and reality. It is a post-foundational prism that allows us to see *El Chavo* as complicit with power *and* as a metaphor of Latin American schooling *and* as a challenge to regional pedagogical imaginaries *and* as a subversive figure *and* as a new kind of literary figure *and* as more of the same *and* . . ., without ever settling for one of these possible Chavos over the others. This book does not attempt to exhaust the analytical and experiential potential of *El Chavo*, but to open it up to further exploration. The task is not to understand, but to proliferate and create, as the Chavos we imagined never cease to argue with, contradict, add layers to, and question not only each other, but the Chavos not yet produced.

El Chavo as rhizomatic curriculum

This rhizomatic Chavo is placed in tension with and against pop culture studies. Pop culture remains suspect in many a critical scholar and/or educator's perspective. Different takes and offshoots of Marxist social theory continue to see pop culture as ideological production that gets in the way of class consciousness by spreading messages that sustain the capitalist order (see e.g. Sum, 2016; Wagnleitner, 2001). The mere facts of reaching the masses and being produced mostly by for-profit ventures mark pop culture with the burning brand of complicity. The absence of a serious study of pop culture in the vast majority of teacher education programs and schools of education also reinforces the notion of it as contamination, a miseducation that gets in the way of "real" culture and knowledge. If anything, pop culture is included as bait; that is, as ways of engaging young students so that they have access points to help them learn the "real" curriculum. Our rhizomatic Chavo—most definitely an integral part of Latin American pop culture—does not seek to aid or hinder the official curriculum, but it *is* the curriculum, that is, a rhizomatic curriculum. We understand both pop culture and curriculum to be themselves impossible—and undesirable—to unify under one set of meanings and interpretations; they bear no single purpose, nor do they carry a universal function. Not only that: neither curriculum nor pop culture can be organized under any hierarchical principle. Not that people have not tried.

The rhizomatic Chavo-curriculum plays with borders in unique ways. It does not ignore them, nor does it erase them, but it does perforate them. When the rhizomatic Chavo-curriculum travels, it is still always Mexican, even when it becomes something else. When watched by a six-year-old Argentine child in 1984, it spoke a foreign version of a familiar language; it presented a strange reality that was at the same time indistinguishable from his own, it came from far away but it *felt* like a close friend. It produced a landscape of *vecindades* that were neither *barrios* nor *villas*,[3] that showed hunger without pain and unequal structural conditions without consequences. A stuck meritocracy in which the moral lessons (e.g., to help each other, to have empathy, to love one another) had no effect as each new day the characters and settings would go back to the way things were and had always been. When watched by an eight-year-old Venezuelan child, the show offered a palpable sense of familiarity, particularly in regard to school and childhood antics. While there was an underlying recognition that the show was from "elsewhere"—one could not ignore the unfamiliar linguistic phrases or the proliferation of Mexican historical facts that were often (incorrectly) recounted to Profesor Jirafales—the recitation of mind-numbing facts to schoolteachers, the ubiquitous tattle-telling of peers and then making up, the childhood crushes, and the invention of games using only the available resources, were comedic and relatable sites of identification.

In his early reflections on globalization, Zygmut Bauman (1998) dedicates a chapter to mobility, specifically to what he saw as the fact that in global times, everyone feels the urge to be on the move. He was clear to make a distinction, however, between tourists and vagabonds. The former travel because they can, as a status symbol as well as a power move that allows them to visit different locales without getting involved. The latter move because they have no other choice; they can never settle. These categorizations also apply to El Chavo's perforations in Latin America: he is a vagabond-tourist, a media product that is made to travel, bringing with him his fabricated misery, an aestheticized refugee that—beyond the intentionality of his creator and the media conglomerate that put him on the move—both comforted and unsettled the televised landscape of Latin American childhoods. There he was, an orphan, earning his keep among the He-Mans, Superfriends, and Strawberry Shortcakes of sedentary afternoons. Already a Spanish speaker, El Chavo was one of the few original voices among an ocean of dubbed awkwardness.

Here, too, we can use Thomas Popkewitz's (2005) conceptual themes of "indigenous foreigner" and "traveling library" as frames through which to consider how *El Chavo del 8* traveled across Latin America. Although Popkewitz's concepts were originally applied to John Dewey as a "conceptual persona" whose notions of the modern self traveled as a model for citizenship around the world, we can relate these ideas to El Chavo in an effort to examine how the show (and its characters) journeyed and intersected with different contexts, producing certain ways of seeing and being, along with particular forms of curriculum and pedagogy. As "indigenous foreigners," the cast of *El Chavo del 8* took on different identities as they landed outside Mexican borders. For example, Don Ramón has become an unlikely staple of street art and t-shirts in contemporary Argentina, standing for the improviser who is continuously hit by crisis and needs all his wits to make the best of terrible conditions. This Don Ramón is as Argentinian as Maradona or Evita, two "characters" who have themselves become indigenized in different locations through multiple cultural appropriations. Relatedly, the series also provided a "traveling library" of ideas about schooling (see Arata and Friedrich in this volume), childhood (See Frangella and Parra-Moreno), daily life (see Colmenares), comedy (see Caraballo), media (see Cabrera and Carbajal Romero, and Treviño Ronzón), and poverty (see Frangella and Parra-Moreno) that became naturalized or indigenous in new cultural contexts. For instance, the quite specific spatial arrangement of the *vecindad* becomes a lens through which to read *conventillos*, a similar yet different housing complex specific to Argentina at the turn of the twentieth century among mostly immigrant communities in the capital city. Thus, the show offered ideas that could integrate dissimilar or conflicting paradigms, whether those were linguistic, educational, politico-discursive, or socio-historical, among others. Never just moving from a supposed "center" (i.e. Mexico) to a periphery (i.e. the rest of Latin America)—and always mediated by specific socio-historical discursive frames—*El Chavo del 8* resonated with viewers in particular ways. Many of the chapters in this volume implicitly articulate the tropes of "indigenous foreigner" and "traveling libraries," providing evidence of how the show traveled across borders to produce, distort, change, incorporate, or indigenize new ideas.

How does this book work?

For Deleuze and Guattari (1987), what a book "means" is not as important as *what it does* or *how it works*. As such, we wanted to create a book that produces difference over sameness, temporality over totality, and multiplicities over meaning and representation. For organizational purposes, the book is split into three sections: Plateau I: El Chavo Encounters Latin American Education and Childhoods, Plateau II: El Chavo's Encounters with Other Latin American Societies and

Cultural Artifacts, and Plateau III: El Chavo's *Media*-ted Encounters. While the content and concepts differ from section to section and chapter to chapter, they are all intricately connected, making sense *through* their relations with each other. The emphasis or purpose of this volume is not necessarily in its individual "elements" or chapters, but rather in the relations of the "in-between" (Coleman & Ringrose, 2013, p. 9), for it is "there" that new things can be produced or imagined.

Relatedly, reading in a Deleuzo-Guattarian vein requires one to "plug in." We view this as a process of arranging, organizing, fitting things together, and/or experimenting. As a reader, the purpose of "plugging in" is to produce something new (about *El Chavo del 8*); to engage in a constant, continuous process of making and unmaking. With that, we ask you to "plug in" wherever you like. Like a nomad, follow what resonates with you. Allow yourself to be *arrested,* where to arrest (if we follow a Derridean chain of significations) means not only to seize, but to stop, halt, or make a brief visit (Jackson & Mazzei, 2011). Pace your brief visits so they are governed less by a sense of totality and more by temporality: where the goal is to play with "temporary meaning[s] that can escape and transform at any moment" (Jackson & Mazzei, 2011, p. 6) and generate something inventive about *El Chavo del 8.*

Plateau I: El Chavo Encounters Latin American Education and Childhoods

Chapter 1 combines a historical approach to Latin American pedagogy with a curricular, post-foundational reading of popular culture to analyze scenes of schooling. Through a close analysis of 23 sketches that take place inside El Chavo's "unique classroom"—"unique" because literally nothing is learned, no progress is made, and no one changes—authors Arata and Friedrich expose the multiple and contradicting ways in which the show's images of schooling simultaneously reinforce the modernist ideals of schools and present some of the specific challenges that schools in Latin America face: schools where the only thing that is ever learned is *how* to be in school, a school that saves no one, and a school that, without being cynical, is not concerned about the future. The significance of the chapter lies in exploring how *El Chavo del 8* allows and provokes questions around what (if anything) constitutes the imaginaries of schooling in Latin America and how the show itself produces such histories and imaginaries among millions of schoolchildren across the continent, a process in which both of the authors are intimately implicated.

By exploring how notions of childhood are presented and constructed through the silver screen, Chapter 2 examines how *El Chavo del 8* participates in the discursive practice and production of childhood. It considers how characters such as El Chavo and his friends are fashioned in relation to dominant discourses about children, and how, in turn, they contribute to a public discourse and imagery that produces knowledge and beliefs about childhood. Author Parra-Moreno stresses the social construction of childhood that emerges from the interconnected images and discourses about children and argues that the cultural production of El Chavo challenges linear and deterministic depictions of children.

Chapter 3 reflects on the popularity of the series in Brazil and considers how the show's theme of childhood has implications for education and the study of childhood itself. Drawing from the work of Homi Bhabha, Arjun Appadurai, and Jacques Derrida, authors Carvalho and Frangella urge readers to view childhood as a liminal space of cultural production that is separated by difference as well as considering the curricular role that the media plays in the production of such images and meanings (Appadurai, 2004; Pinar, 2011). According to the authors, the images of children created in *El Chavo del 8* contribute to the construction of imagined worlds, often fantastical in nature, and may even dislocate existing conceptions of childhood by allowing imagination to be seen as a social practice through which childhood might be perceived differently. Relatedly, by breaking binaries which mark as fixed the place of the child and the adult in educational processes, this chapter discusses how the show pushes its viewers to denaturalize their previous, often "essentialized", images of childhood.

Plateau II: El Chavo's Encounters With Other Latin American Societies and Cultural Artifacts

Chapter 4 traces the parallelism and equivalencies between *El Chavo del 8* and the tradition of the picaresque novel, a literary form known for its (often grim) portrayals of social and urban realities. To do this, Aguasaco offers a comparative reading of the representations of childhood, schooling, hunger, and social structure in *El Chavo del 8* along with four picaresque novels: *El Lazarillo de Tormes* (1554), *El Buscón* (1626), *El Periquillo Sarniento* (1819), and *La vida inútil de Pito Pérez* (1938). By comparing the residual and emerging ideological elements of the show and the four picaresque novels, this chapter demonstrates how popular culture can be understood as an articulatory form of processing, producing, and reproducing symbolic and ideological content that uses new formats and technologies.

As implied by the title, Chapter 5 examines what happens when two international and intergenerational cultural icons, Chespirito's El Chavo and Pepito, a mischievous personage in many Latin American jokes and tales, cross borders. Through a narrative inquiry of reflections about El Chavo and Pepito gathered from 12 members of various generations among family and friends across the Caribbean and Latin America, author Caraballo explores how these cultural icons were experienced in similar and dissimilar ways and how they were framed as sites of possibility for rethinking static notions of identities, cultures, and citizenship with/in immigrant and minoritized communities. Exploring the similarities and differences among audience recollections of these figures invites a deliberate engagement with some of the political and socioeconomic narratives that inform current social discourses and power relations in Latin American societies.

Among the wide array of Latin American societies and cultural artifacts, Chapter 6 traces *El Chavo del 8's* encounters within a specific spatio-temporal context: Venezuela during the 1980s and 1990s. Guided by the theoretical possibilities afforded by the Affective Turn, the chapter examines how the production and proliferation of particular affects and structures of affective attachments may have helped pave the way for the series to become an important site of identification and recognition in the country of the author's youth. Specifically, author Colmenares posits that the show's magnetism can be attributed to its capacity to forge an "intimate public" (Berlant, 2008) among its viewers—an aesthetic and affective structure that binds individuals together through emotional ties, promoting a feeling of belonging or being in common. The significance of this chapter lies in examining how a mass-mediated "text" such as *El Chavo del 8* can create reciprocal, emotional ties among strangers in different parts of the world and produce a space where people can come together to dream, work, or experiment alternative attachments, forms of adjustment, or modes of relationality in a world that, like Venezuela in the 1980s and 1990s, continues to fray at the seams.

Plateau III: El Chavo's *Media*-ted Encounters

To conclude the volume, the final two chapters explore the intricate relationship(s) between *El Chavo del 8* and the media. More specifically, Chapter 7 explores the political links between television content and the production of identifications and social representations of schooling. Using the theoretical notions of interpellation and identification and the work of Roland Barthes, Slavoj Žižek, Jacques Rancière and Jacques Derrida, author Treviño Ronzón argues how Chespirito provided an engaging short narrative about a school in a society marked by poverty, exclusion, and violence, producing an interesting and complex process of happy "popular identification." The chapter explores how social representations and identifications generated in TV shows such as *El Chavo del 8* should be problematized to make possible new explanations about the making of identities, social binding, social equality, and in particular, the dynamics of politization around the media. Specifically, Treviño Ronzón focuses on the production and circulation of social stereotypes, the explicit and implicit violence deployed by and through the show's characters, and the ways in which the show tells "Stories" about society, social agents, and their relations with the state.

Working from the assumption that culture is inherently embedded in media education, Chapter 8 analyzes the series to argue that media(ted) education happens "*sin querer queriendo*" (unintentionally). According to authors Cabrera and Carbajal Romero, media(ted) education occurs in an *intermezzo* space, between the televised (transmitted) content and the process of subject identification (Deleuze and Guattari, 1987). Using tools from hermeneutic analysis to understand the processes of subjectivation and culture through media production, this chapter links some of the various authors and concepts that shape a post-foundational view of media theory and cultural studies, including but not limited to: the urban-popular, cultural frames, and mass media. Cabrera and Carbajal Romero explore how the intercultural, affective and symbolic reconstruction evoked by the figure of El Chavo combines different ways of living everyday urban life and includes several educational, political, comical, and aesthetic dimensions, all "*sin querer queriendo*" [without meaning to].

Conclusion

Taken together, these chapters trace the ways in which *El Chavo del 8* has traveled across borders, impacting the linguistic, discursive, educational, sociocultural, and affective practices across a wide array of people, time(s) and space(s). In line with Brian Massumi's (1987) claim that a book is an open system that does not pretend to have the final word, we make the same assertion here and heed his advice in approaching this volume. A book, like a record, is meant to be "played":

> When you buy a record, there are always cuts that leave you cold. You skip them. You don't approach a record as a closed book that you have to take or leave. Other cuts you may listen to over and over again. They follow you. You find yourself humming them under your breath as you go about your daily business (1987, p. xiv).

We encourage you to approach this book in the same way. Regardless of where your nomadic forays take you in this volume (St. Pierre, 2000), we conclude by inviting you to watch *El Chavo del 8* for yourself and feel the affective resonances it has to offer.

Notes

1 This had to be clarified after the show moved from Channel 8 to Televisa; a new reason had to be given for "del 8" so the show could still be called "El Chavo del 8."

2 This was, of course, extremely uneven across the continent, making it impossible to make such generalizations for differences between, for example, Bolivia and Mexico.

3 *Villas* in Argentina are urban slums, the equivalent to Brazilian *favelas*.

References

Aguasaco, C. (2014). *¡No contaban con mi astucia! México: Parodia, nación y sujeto en la serie de El Chapulín Colorado* (1st edn). Mexico DF: Universidad Nacional Autónoma de México.

Appadurai, A. (2004). *Dimensões culturais da globalização: A modernidade sem peias.* Lisbon: Teorema.

Bauman, Z. (1998). *Globalization: The human consequences.* New York: Columbia University Press.

Berger, P. L., Berger, B., and Kellner, H. (1974). *The homeless mind.* Vintage Books.

Berlant, L. (2008). *The female complaint: The unfinished business of sentimentality in American culture.* Duke University Press.

Chin, E. (2016). *My life with things: The consumer diaries.* Durham: Duke University Press Books.

Coleman, R., and Ringrose, J. (Eds). (2013). *Deleuze and research methodologies. Data across multiple perspectives.* Edinburgh: Edinburgh University Press.

Deleuze, G., and Guattari, F. (1987). *A thousand plateaus: Capitalism and schizophrenia.* Minneapolis: University of Minnesota Press.

Jackson, A. Y., and Mazzei, L. A. (2011). *Thinking with theory in qualitative research: Viewing data across multiple perspectives.* London: Routledge.

Massumi, B. (1987). Translator's note: Pleasures of philosophy. In G. Deleuze & F. Guattari, *A Thousand Plateaus: Capitalism and Schizophrenia*. Minneapolis: University of Minnesota Press.

Pinar, W. (2011). *What is curriculum theory?* New York: Routledge.

Popkewitz, T. (Ed.). (2005). *Inventing the modern self and John Dewey: Modernities and the traveling of pragmatism in education*. Basingstoke: Palgrave Macmillan.

Puiggrós, A. (1996). "Presencias y ausencias en la historiografía pedagógica latinoamericana." In H. R. Cucuzza (Ed.). *Historia de la Educación en Debate* (pp. 91–123). Buenos Aires: Miño y Dávila.

St. Pierre, E. (2000). "Nomadic Inquiry in the Smooth Spaces of the Field: A Preface." In E. St. Pierre & W. Pillow (Eds). *Working the ruins: Feminist poststructural theory and methods in education*. New York: Routledge.

Sum, N.-L. (2016). "The Makings of Subaltern Subjects: Embodiment, Contradictory Consciousness, and Re-hegemonization of the Diaosi in China." *Globalizations*, 1–15.

Villamil, J. (2012). *El sexenio de Televisa: Conjuras del poder mediático*. Mexico DF: Penguin Random House Grupo Editorial México.

Wagnleitner, R. (2001). "'No Commodity Is Quite so Strange as This Thing Called Cultural Exchange': The Foreign Politics of American Pop Culture Hegemony." *Amerikastudien/American Studies*, *46*(3), 443–470.

Plateau I

El Chavo Encounters Latin American Education and Childhoods

1

Schooling, Pedagogical Imaginaries, and Latin American Childhoods in *El Chavo del 8*

Nicolás Arata and Daniel Friedrich

The different ways of living and going through childhood in Latin America have constituted a deep pool of experiences in which—as explained by Susana Sosenski and Elena Jackson Albarrán (2012)—one can find plenty of images linked to the economic roles of boys and girls (as child workers or consumers), mentions of *dispositifs* of institutionalization and imprisonment (consequences of ideas and practices that identify the child as savage, delinquent, or dangerous), and histories of the ways in which childhoods participated in organized citizenry, armed movements, or school initiatives, among others. According to the authors, the specificity of these experiences does not lead one to conclude that Latin American childhoods are inherently different than those of other regions. However, the political, economic, and cultural transformations that impose a particular rhythm on relations between state, families, market, social institutions, and labor in Latin America configure historical situations in which the ways of living through childhood acquire distinctive characteristics. Similarly, this does not imply one can define a single Latin America, or a unified Latin American childhood: each one of these concepts contains multitudes.

One way to approach the study of representations of Latin American childhoods is by focusing on comics and children's television shows. The characters that inhabit the worlds of *Mafalda*, *El Chavo* or *A Turma da Mônica* are globally recognized and operate as cultural ambassadors who, through their stories, bring images close and afar of different ways in which children in the continent live their first few years. These narratives—elaborated by adults—spell out the experiences and issues faced by children in regard to multiple topics: family relations, cohabitation practices (both public and private), ways of appropriating public space, and of participating in the labor force, among others.

But how do these stories nurture and take part in Latin American popular culture? Stuart Hall (1998), in his deconstruction of the concept of Popular Culture, highlights the difficulties of working—from a critical perspective—on an area of the social weave that, far from being transparent, presents a variety of nuances. Are popular cultures such because they contain a traditional, oral, premodern quality, or are they the result of a homogenizing effect of the cultural industry? Could one term "popular," phenomena emerging from such different locations as urban or rural spaces, the indigenous world, or the workers' culture? (García Canclini, 1987). Yet the fact that it is a slippery and polysemic concept does not detract from its value. According to Hall, popular culture is a site for the struggle of meanings and a platform from which potentialities can be deployed. The mass appeal inherent in the production of a popular culture is always already suspect of being complicit with power. In other words:

> If the forms and relationships on which participation in this sort of commercially provided "culture" depend are purely manipulative and debased, then the people who consume and enjoy them must either be themselves debased by these activities or else living in a permanent state of "false consciousness." They must be "cultural dopes" who can't tell that what they are being fed is an updated form of the opium of the people (Hall, 1998, p. 446).

However, as Hall indicates, this view of "the people" that embodies the popular as a passive subject facing the forces of capitalism is not only patronizing and insulting, but deeply anti-socialist. Neither does this imply maintaining a depoliticized view of popular culture. On the contrary, Hall demands of us a complex reading, one that respects the tensions inherent in the processes of production of subjectivities mediated by mass media. Giroux and Simon (1989) add to Hall's perspective by including the need to consider the issue of pleasure in any critical analysis of popular culture as a site of public pedagogy.

These and other dilemmas lead us to view *El Chavo* as a cultural artifact through which to examine the relations between pedagogical imaginaries, schooling, and Latin American childhoods. As authors who experienced our childhoods in the 1980s, we were affected by the show created by Roberto Gómez Bolaños and touched by the images of Latin American popular culture spread by the series. During our early years in Argentina, we learned that in Mexico "chavo" was another term for kid, that *vecindades* (housing projects) were colorful spaces full of adventure, and that *tortas de jamón* were a valued commodity, not just a disgusting combination of meat and chocolate.[1] We also understood that any offense—intentional or not—towards an adult could *naturally* lead to a *zape* [slap].

These images (and our own reading of them, forever full of memories) can be treated as sources through which to interpret how a society produces, circulates, and appropriates—through mass media—different imaginaries of schooling. By focusing on 23 sketches situated in the classroom, this chapter analyzes the ways in which a specific representation of childhood is framed within schooling, exploring how a comedy show—in which the basic premise is a group of adults dressed as children—can be tackled as a source to consider the process of the coming into being of Latin American schooling. How does this TV show construct a representation of childhood? What do these sets of—apparently absurd—images and situations say about the imaginaries of schooling in Latin America being produced?

That old sacred cow

During the 1970s and 1980s, Latin American schools were the target of strong accusations, arguments, and debate. Reproductivist thought explained how schools had been programmed to legitimate class inequalities. The ideas of Bourdieu, Passeron, and Baudelot spread like wildfire through the continent's universities challenging the image of school as "social equalizer" as conceived by nineteenth-century liberals. In its place, reproductivist theorists counterposed the picture of a complex machinery configured to reinforce the social origins of each student.

At that time, popular education movements multiplied Freirean perspectives that questioned some of the foundations of modern pedagogy, yet without renouncing the pedagogical optimism that motivated it. A different position consolidated itself in the classrooms of the International Education Center in Cuernavaca, where calls to de-school society issued by theologian Ivan Illich—who accused formal education of being the "sacred cow" of capitalism and a tool to perpetuate power structures—placed in front of schools a question about their legitimacy that, until then, few people had dared to ask. Paradoxically, while all these critiques were being explored, Latin American school systems reached coverage like never before. The modern promise of access to reading, writing, and calculus, among other cultural goods, coexisted with severe warnings about school's decay and corruptive effects.

We refer to the school because this institution is the common denominator that holds all the fictional characters we mentioned together. Each character comes from a different social stratum: Mafalda belongs to a "typical" middle-class family in the city, lives in an apartment building, and displays certain cultural capital, whereas El Chavo is an orphan who seems to live in a barrel, in a *vecindad* always flitting between the possibility of affording the rent or falling apart. Yet all of them go to school.

El Chavo's schooling takes place in the 1970s, in an overpopulated Mexico City which two years earlier had been thrown into upheaval by the Tlatelolco massacre, and which by the beginning of the decade had hosted the FIFA World Cup. The ways of living *Mexicanity*, in a city in which each gesture invites an archeological dig in search of the deepest cultural codes, can be read through a double prism. On the one hand, there are elements that contributed to the Latin Americanization of the national culture: Mexico had always been the *casa grande* [big home] of Latin Americans, accentuated by the waves of exiles fleeing dictatorships in the Southern Cone, seeking refuge in Aztec lands. On the other hand, the culture was being Americanized, through both economic and technologic dependence, as well as the proliferation of US-based mass media.

Framed by these processes, we claim that El Chavo's richness is tied to the fact that the representation of Mexicanity specific to *that* social class, in *that* city, during *that* decade, taking part in *those* processes, is juxtaposed to other elements that led us, the authors of this chapter—together with millions of people in the whole continent, through the generations up to the present day—to adopt Chavo into our most intimate friendship circle, to run home from school so that we wouldn't miss a minute of the episode (that we had already seen a million times!), and to repeat the catchphrases in our everyday conversations even though we did not always understand them in the same ways.

From the multiple situations, characters, and cultural sites offered by the show, our analysis attempts to account for that juxtaposition as an integral aspect in the scenes of schooling in *El Chavo del 8*. In that school, Profesor Jirafales leads a classroom that, through the series, tackles different subjects. We approach El Chavo's school identifying how each subject matter is treated; the ways in which the students, parents, and teachers are represented; the explicit, hidden, and absent curricula; the sites of authority and knowledge; and the role of schooling in ideas of social progress and change. Our goal is to identify in these registers of school life the modes in which certain representations of childhood become entangled with the pedagogical imaginaries available at the time.

One day in El Chavo's school

The classroom as a stage

The classroom, as usual, has three visible walls and an implicit one, offering a gaze that assumes itself as the totality of what takes place in that space, guided by that which the camera chooses to focus on. The teacher's desk is at the front of the classroom, situated on a slightly elevated platform that operates symbolically as a barrier to remind everyone that he who possesses knowledge is a bit closer to heaven than the rest. The students' desks are distributed in rows of three, in an improbably small classroom for what is assumed to be an urban school in the second half of the twentieth century. The walls with drawn windows are covered in educational posters, maps of Mexico, and paintings with the faces of the founding fathers. Although the sketches were filmed over more than 15 years, the design of the classroom space, the technology utilized (chalkboards), the furniture, and the decorations remained basically the same, following one of the show's themes.

During history class[2, 3]

Mexico, a nation exalted by poet López Velarde as "impeccable and diamantine" (1988), becomes in *El Chavo del 8* a puzzle to be put together. The history class builds bridges to the distant, premodern past. As is the case in the rest of the school scenes, orality dominates over the written culture. As opposed to the scenes that take place in the *vecindad*, where the characters have more freedom of movement—allowing for more physical humor—every gag in the classroom is grounded in verbal comedy; specifically in what in Mexican popular language is known as "albur" [pun, double entendre]. Marco Antonio Montes de Oca refers to *albur* as the "verbal jazz of Mexicans." Far from referring to a degraded form of language, *albur* is an ingenious and creative strategy for maintaining a conversation in which "the one with the best *albur* is a potential leader" (Montes de Oca, quoted in Mejía Prieto, 1985). Thus, the language used in *El Chavo del 8* carries with it a plebeian character that challenges the cultivated knowledge—or what Apple (1996) calls the official knowledge—without confronting it in the open. It undermines the teacher's authority appealing to linguistic turns that step outside the script and which contradict and ridicule him. What we are highlighting here is that in El Chavo's classroom, one finds not any discourse, but popular discourse.

The relationship to history is extemporaneous to the class. There are no organic connections between past, present, and future, neither in the instructor's questions, in the students' answers, nor in the lack of attempts to take into account the complexity of historical processes. History is external to the group, a mere partition of time. When Prof. Jirafales mentions the history test, Chilindrina demands that there be "no cheating," clarifying that "you always ask us about things from before we were even born." Far from correcting her, the teacher (perhaps unintentionally?) legitimates the student's concern, replying: "Well, that's what [history] is about!"

The history lesson follows a capricious sequence, without chronology or periodization, centered on the "founding figures" of Mexican history. The series of questions asked by Prof. Jirafales starts with the priest Hidalgo in 1810, followed by the execution of Emperor Maximiliano in 1862, then Moctezuma's death in 1520, to end with the takeover of Tehuacán by Francisco Villa in 1911. There is nothing weaving these events together, nor are there any follow-up questions for or by students; only disconnected milestones between heroes, emperors, and revolutionaries that never talk to each other.

The students' answers denote a literal reading of historical events. Thus, for Ñoño, the reason behind the priest Hidalgo's "Grito de Dolores" (the proclamation that initiated the process of Mexican independence against Spain in the city of Dolores, literally translatable as the "cry of pain"), was that he was "very sick." In the same way, Chavo affirms that Hidalgo—also known as the Father of Mexico—was married to Spain, since "Spain is the Motherland." When Prof. Jirafales attempts to redirect the class by asking about the main obstacles that the insurgents had to deal with, Quico replies: "traffic lights," referring to the large avenue (Avenue of the Insurgents) that crosses the city from north to south. Without changing the format of questions and answers, the teacher asks where the Declaration of Independence was signed, to which Popis—always the eager student—retorts: "on the bottom of the sheet."

Finally, there are no references to the events that shook Mexico during those years; El Chavo's classroom windows did not face Plaza de las Tres Culturas, where hundreds of students were massacred in 1968, a few years before the show started filming. As opposed to Mafalda, El Chavo does not meddle in explicitly ideological affairs, past or present. What it does deal with is the tight relationship between class, written culture, and moral education. The teacher claims: "Children, take your books and listen closely. While you hold a book in your hands, you will be worthy people [*gente de provecho*], good people." Thinking about the arbitrary content of the history class, it is not likely that the textbook would make students "worthy people." What makes them potentially worthy is their mere presence in school, where moral lessons are communicated through curricular content.

Adults go to school as well[4]

The classroom walls are not usually porous. Yet at specific times, subjects that tend to stay outside are invited in, interrupting the normal functioning of school. Parents are rarely seen in the classroom, beyond the initial moment in which they leave their children under the teacher's care. But "today" is different. Parents are present in school to verify that their children have learned something. Three of the students' parents have been invited by the institution to observe, provided that they "abstain from cheating." The invitation was issued to the parents of the three students who failed: Chilindrina, Quico, and Chavo. Of the three, only two bring accompanying adults. Chavo, as an orphan, has no one to mediate between himself and the teacher.

The first part of the exam is on drawing, an unusual subject to fail (and to "cheat") on. The three students seem to work hard on their drawings. Chavo is obsessed with perfecting his work, to the point of forcing the teacher to remove it from his hands. Chilindrina draws a typical doodle, and is awarded the "six" needed to pass.[5] Quico builds a story around a sheet of paper. He turns in a blank sheet, yet full of narrative: it is a cow eating grass; the grass has been eaten and the cow left. The teacher seems to have no other option than to grant him a "six." Chavo, on the other hand, presents an abstract colorful drawing, which the teacher accepts with disgust. "It's a chiforíncula," explains the proud student. "And what is that?," asks the teacher. "It's something I made up. It came out identical, right?" In the country of alebrijes,[6] the "chiforíncula" is an original creation. The teacher, trying just to move onto the next topic, gives him another six. Yet Chavo resists his indifference: "Why? It is a perfect copy [of what I imagined]. You should give me a ten!" If schools will test (and fail) the students' artistic potential, Chavo demands that originality not be sanctioned.

The second part of the exam focuses on Language Arts. But language, that living matter that feeds most jokes and class interactions, is torn into decontextualized, meaningless questions. When the teacher asks Chilindrina for two pronouns, the student replies anxiously: "Who, me?" Proving that the traditional pedagogy that removes all possibility of sense-making is arbitrary and does not always place students at a disadvantage, the teacher gives her a ten: the two pronouns have been enunciated.

When Quico cannot answer a question, his mother is ashamed of his ignorance, and when Chilindrina replies correctly, Don Ramón (her father) shows pride, which leads to a confrontation between the parents. For them what is at stake is their own value as educators, even though neither of them seems to have any clue what the exam is about. Curiously, no one questions the teacher as to why the students do not know the content. Here there are only two possible explanations: that the children are not smart enough, or that they have not studied. Prof. Jirafales ends the class completely frustrated, but not necessarily with himself. His frustration is mixed with pain, stating that he would have preferred if no one had failed. At least—he says—"only one deserved the donkey ears": Don Ramón, who is standing in a corner, crying and humiliated, dons the damning symbol.

Don Ramón's revenge[7]

School can also be the site of inversions, where he who previously donned the donkey's ears is presented with the possibility of assuming the role of the teacher. The porousness of the classroom walls has the potential to translate into a positional inversion. When Prof. Jirafales needs to step out of the classroom (his beloved Doña Florinda is waiting outside the school), he leaves Don Ramón—who is there precisely to escape Doña Florinda—in charge of the class. "Of course, I was also once a teacher," Don Ramón reassures an incredulous Prof. Jirafales, "I was a foreman [*Maestro de obras*]."[8] The laughter in the background and the teacher's facial expression combine to denote what is perhaps obvious to the viewers: that the teacher's and the worker's knowledge are not interchangeable.

Don Ramón represents, during the show, the lowest of social hierarchies in the *vecindad*, barely above Chavo. His extremely thin body triggers one joke after the other linked to pain and hunger. Don Ramón always owes rent, never holds a steady job, and tends to demonstrate very little cultural capital (his response to most gags tends to be a physical reaction). He embodies, as Doña Florinda calls him, "la chusma" [the rabble], with whom it is better not to mix. This is why, leaving him in charge of the classroom can be seen as a somewhat subversive act.[9]

School produces *habitus*, standardizes reactions, and allots roles. Don Ramón *knows* that first one needs to have order, and begins class with what seems like traditional pedagogy, similar to that he experienced in his years as a student. Questions and answers, grades and authority. Under that general structure, a fundamental shift takes place, a reversibility that makes visible school's potentiality. Don Ramón asks: "Popis, what costs more, a kilo of lettuce or a kilo of tomatoes?" "I don't know," replies the surprised student. "Flunked. Ñoño, what is worth more points, a touchdown or a field goal?" "I don't know." "Flunked."

This time, the laughter track is more ambiguous. Is the joke coming from an educated spectator finding it ridiculous that such mundane questions become legitimized content in school? Or how easy it is to expose school's arbitrariness? What happens when ignorance of the everyday leaves traditionally better-positioned students at a disadvantage? Both laughter and ambiguity return in a unique moment for the show: Godínez, who is always lost in school and obsessed with sports, responds correctly for the first (and only) time when Don Ramón asks a question about poker, and the interim teacher gives the student an approving wink. Godínez's whole posture changes, at least for a few seconds, as he finally belongs to the classroom.

When Prof. Jirafales returns to the classroom, he meets a group of students fascinated with Don Ramón the teacher. Suspicious at first, he observes how the man who had been previously crowned with donkey ears teaches children to identify poisonous bottles by the skull and crossbones symbol, and not to put their fingers in the electrical outlet. These life-and-death subjects had never been considered worthy (or had at least been underestimated) by the official curriculum. Nevertheless, here they were, being taught by the least expected of instructors. Prof. Jirafales applauds a successful Don Ramón; valued and appreciated, Don Ramón's body straightens up, proud of the acknowledgment of a school that most likely had expelled him years earlier.

The economy of those who have little[10]

The economics lesson begins with students and teacher occupying their usual places. There are eleven children (of which only four will actively participate, the rest little more than bodies filling up the space) dressed in colorful clothes that do not represent the everyday wear typical of that age. All but one, set apart by his torn outfit set, and perhaps the effects of the economy that the teacher would soon tackle.

Prof. Jirafales starts by stating that everyone must have heard about economics, "especially lately." And starting from this assumption, something different happens: before the lesson begins, the teacher asks the students whether they would like to talk about this issue. The interest expressed by them legitimates the lesson plan.

The first definition provided by Profesor Jirafales makes it explicit that the economics that he will discuss in class, and those producing differences among the children are not necessarily the same: "Economics is the science that studies the systems that need to be applied to avoid unnecessary expenses." At least, that seems to be the teacher's intent. However, almost immediately this is subverted by one of the students. If the pedagogue wanted to explain "unnecessary expenses" in a decontextualized and impersonal manner, Godínez proposes an example: "Ñoño's house . . . because it has seven bedrooms, four dining rooms, eight living rooms, and all of that, only for Ñoño and his dad!" Chilindrina jokes that it is not a waste, given Ñoño's size, which of course causes him to protest, asking the teacher to intervene. Yet the instructor, teasing perhaps one of the lessons he will impart, comments that he does not consider it an insult to talk about waste in those terms. The economics class begins to turn into a morals class.

The teacher covers topics such as savings and inflation, in order to get to the final lesson. "Oh, I get it, what happens is that you all are jealous of me, because I'm a much more expensive child than the rest of you," cries Ñoño, the landlord's son. "Even if it pains you, even if it pains you. Especially you, Chavo. Because you can't deny that you are a very cheap boy." "Yeah, well, that's true," accepts a resigned Chavo. Immediately, the teacher intervenes: "Ñoño, if what you are saying is that Chavo is a poor kid, you should know that he will soon cease to be one." To which Chavo retorts: "When?" "Well, some day. Two things are necessary. First, that you study a lot." Prof. Jirafales bets on meritocracy, navigating the uncomfortable tension between school's optimism and resignation about that which he has no control. At least for now. This bet makes visible the contradiction grounding the show, one that runs parallel to the contradiction of the non-televised school. Chavo will never cease to be poor because that is a characteristic that is *fixed* to his identity. His behavior will not change as a result of content learned, despite the teacher's belief that school offers a social and moral-climbing ladder. If bodies, spaces, and school forms remain unchanged along the show, why would Chavo's manifest destiny change, *thanks to an immutable school?*

Chilindrina accuses Chavo of not studying. Chavo refutes this, explaining that his problem is a lack of memory. The teacher jumps in:

> "Exactly, you need to keep things in your memory, and keeping is like saving, and that is the second condition. The second condition is that you must save." "Save what?" "Money!" "What money?" "Chavo, you must have held a coin in your pocket at one point, right?" "Yeah, but my pocket had a hole! But you are right, Jaimito the mailman once said the same thing to me . . . that people can save things in their memory. Meaning, that it's good to save, but not always." "What do you mean 'not always', Chavo?" "Well, only when one saves good things. Because for example if one has garbage, it is not good to save it . . . And

> the same with what ones saves in one's memory. Because for example if a kid comes and hits you really hard, it is not good to save it in your memory, because then you remember it and it hurts again. [Queue emotive music]. Instead, when something nice happens to you, it is good to save that, because then you remember it, and you feel good again. And happiness is feeling good. That means that it is convenient to save in your memory good things, in case at some point you get hit with happiness inflation."

The pedagogical intervention inverts the usual order; the student becomes the teacher. Furthermore, a class that began with the teacher expounding the importance of learning about economics, wraps up with the student most directly affected by the economic system reflecting on the role of memory to achieve individual happiness, beyond any structural condition. We cannot but consider the contradictions inherent to visions about poverty emerging outside of it, and the moralizing discourse that leaves everything as it is. At the same time, school is the place where Godínez, a student rarely positioned as a character with knowledge, interrupts the lesson to highlight the impossibility of depersonalizing the experience of schooling, by questioning the injustices woven through his classmates' lives.

The anti-progress class

If history refers to a series of events disconnected from the present, the future does not introduce the image of an alternative temporal horizon. The idea of progress is foundational for modern schooling; it is part of its original DNA. Learning is change, and change is the result of going through schooling across time. Assignments, grades, assessments, and curriculum itself—in its etymological genealogy linked to *currere*, career, racetrack, transit (Pinar, 2011)—are all concepts that are dependent on the time variable.

Most TV shows centered on schooling are marked by the passing of time. In series such as *Glee*, *Beverly Hills 90210* or *Señorita Maestra*, each season is regulated by the school calendar: the beginning of the school year, testing weeks, graduation, vacations. The content of the lesson never repeats itself, as it is assumed that what was covered in the past has been learned. Both students and teachers demonstrate a certain memory related to past events, making references to stories and themes that are familiar to the audience. In other words, going through school denotes growth and progress; it produces marks in the bodies and minds of those who inhabit the classroom.

Some TV shows, such as *The Simpsons* or *South Park*, are the exception to this rule. In these series characters show no growth, and through the decades, remain in the same grade, with the same teachers, learning the same lessons, and demonstrating in each episode the absence of individual change. Obviously, the big difference lies in the fact that both *The Simpsons* and *South Park* are animated shows, thus characters do not need to grow up to account for the physical changes that are evident in the young actors working on *Glee* or *Beverly Hills 90210*.[11]

El Chavo avoids this issue by means of its own premise: adults dressed as children, for whom aging is not tied to change. In El Chavo's school, notions of progress and growth are arrested. What kind of school is one that does not change, under which no marks are produced in the academic biographies of its students and where the teacher does not modify its pedagogical approach facing this absence of change? Is there something particularly Latin American in this idea of schooling as absurd?

The absence of progress in El Chavo's scenes of schooling invites further interrogation. First, let us review the practices that remain constant despite the absence of cognitive change. For example, although there is no evidence of learning or progress from lesson to lesson, there is talk of passing or failing exams, of good and bad grades, or the value of a good education. However, what does it mean to pass a test, get a ten, or to be an educated person, in a context in which "the next day" everything returns to the way it was? In the context of a television show where repetition for comedic purposes is part of the genre, this may seem like a forced question. Yet if one were to place this question in the broader arena of Latin American schooling, it would acquire a different tone, getting closer to the idea of school as simulacrum (McLaren, 1988) than to the idealized modern school. This is the school of the "what if," the school of containment instead of progress, a school in which what is learned is—precisely—to be in school. In El Chavo's school, children rarely know anything about the curriculum, their answers are almost always wrong in relation to what the teacher is trying to teach. But every child knows exactly how to behave in school, who the source of authority and knowledge is and what transgressions are acceptable. What is simulated is learning of the prescribed curriculum, while the hidden curriculum is re-inscribed in each body by daily repetition.

In terms of the purpose of this representation of schooling, we propose that the absence of progress in El Chavo's school opens the door to a different relationship between humor and affect. In TV shows or films where characters go to school, the possibility of getting stuck or dropping out tends to be presented as a problematic situation, acting as a trigger for drama or for efforts to recover what has been lost. In El Chavo's school, all the characters will inevitably stay in school, which disables the negative relationship with educational failure. When a character is punished, or receives a bad grade, the reaction is paradoxically the same as when Chavo is finally valued and granted a good grade: the tenderness produced by knowing that, no matter what, beyond the happiness or sadness expressed by the students, it is merely a temporary emotion and everything will go back to "normal" when children go back to class next episode. It is, maybe, a tenderness that borders on cynicism, since it allows us, the viewers, to create a distance from the consequences of schooling as simulacrum, especially as they relate to the population Chavo is representing.

Echoes, resonances, and specters of schooling

The question of the tension between the mission of schooling and the absence of any notion of progress or growth through time can be juxtaposed with questions about what is specifically Latin American in the representation of the classroom presented by the show, by considering the echoes or resonances generated by the image of Chavo and his friends in *that* classroom on a multigenerational and transregional audience. Given that we are talking about a group of actors averaging 40–50 years of age when the series is being filmed, what school are they performing? Was it the school of their childhoods in the 1930s and 1940s, or the school contemporaneous to the production of the series? Are we facing the institution inhabited by Gómez Bolaños—what he himself described as a "typical" middle-class Mexican school—or that which a poor child like Chavo would have attended, one that responds to the imaginary of schools for the poor?

A possible answer that accounts for the success of the show lies in the idea of a *collage*, conceived as a set of discursive practices that combines, superimposes, contradicts, and puts in tension a diverse array of ideas of schooling, forming a grid that allows almost any audience to establish an affective link with that image of the classroom. If Godínez's absurd answer to a question about the Aztec people refers to Mexico City's soccer stadium, the Estadio Azteca (something that a ten-year-old Argentinian child would not necessarily be aware of), the reaction to that situation is tied to the different subjectivities that the institution of schooling produces in those who attended it. Although the apple is no longer the preferred way to gain the teacher's favor—if indeed it ever was—the fruit resonates among the practices that students and teachers recognize as part of the logic of schooling. The specter of school dropouts—associated with an idea of failed citizenship that *El Chavo* could only aspire to but never reach—mobilizes affects and bodies not only as a specific experience lived by part of the audience, but also as an echo of the potentiality inherent in the very structure of schooling. In this sense, El Chavo's school is the Mexican school of the 1970s, while simultaneously embodying contemporary Latin American schooling.

Mexico City's chronicler Carlos Monsiváis (2007) used to say that in Mexico City, jokes are repeated so often that they become institutions. Through gags that we as spectators could predict, *El Chavo* became an institution of Latin American childhood. Thus, it is not the show's veracity that turns it into an historical source, but the ways in which Gómez Bolaños's creation *produced* history in millions of children's and adults' lives across the continent. It was not just the jokes that were repeated from episode to episode, but the distribution of school space, the content to be taught, resistance to learning, relationships between authority and power, and the ways to subvert them. This reiteration throughout two decades of filming and almost three decades of reruns injected into Latin American pop culture a complex imaginary of schooling, one that is contradictory and resistant to any univocal interpretation. If the dominant products of the Anglo-global pop culture tend to reinforce the idea of school as a meritocratic space, *El Chavo* presents that notion as just one of many unresolved possibilities. Whereas one day school rewards the hardest working, the next day the pedagogical model is completely arbitrary. It is precisely the indifference towards this fact by all who inhabit it what constitutes its "normalcy."

Notes

1 In Argentina, "torta" means cake, while in Mexico, it refers to a sandwich.
2 It is difficult to pinpoint episodes in which one can find the cited dialogue. Most sketches were filmed several times over the years, repeating jokes with minor modifications. The notes indicate the title and year of the sketches as mentioned on the websites we consulted.
3 *El Examen* (1975).
4 *Examen con los padres* (1976).
5 Grades go from one to ten, six being the minimum grade needed to pass.
6 Alebrijes are imaginary figures composed on papier machè or carved in wood.
7 *Don Ramón en la escuela* (1975).
8 This is a play of words between teacher [*maestro*] and foreman [*maestro de obras*].

9 Don Ramón's face has become a cultural icon, featuring on t-shirts, posters, and memes; he has turned into something of a Latin American anti-hero.

10 *Clase de economía* (1990).

11 Weldon (2013) explains the impossibility of growth for characters in comics or cartoons, such as Superman, given their huge value as reproducible merchandising property. As soon as those characters begin to age, the end of profit becomes palpable. A rheumatic Superman does not seem to be a viable product.

References

Apple, M. W. (1996). *El conocimiento oficial: la educación democrática en una era conservadora*. Editorial Paidós.

García Canclini, N. (1987). Ni folklórico ni masivo ¿qué es lo popular? *Revista Diálogos de la Comunicación*. Lima: Federación Latinoamericana de Facultades. No. 17.

Giroux, H. A., and Simon, R. (1989). "Popular Culture as Pedagogy of Pleasure and Meaning." In *Popular culture: Schooling and everyday life* (pp. 1–30). Praeger.

Hall, S. (1998). "Notes on Deconstructing 'the Popular.'" In J. Storey (Coord.), *Cultural theory and popular culture: A reader* (pp. 442–453). London: Pearson Education.

López Velarde, R. (1988). *La suave patria y otros poemas*. México: Fondo de Cultura Económica.

McLaren, P. (1988). "Schooling the Postmodern Body: Critical Pedagogy and the Politics of Enfleshment." *The Journal of Education*, *170*(3), 53–83.

Mejía Prieto, J. (1985). *Albures y refranes de México*. México: Panorama.

Monsiváis, C. (2007). *Las alusiones perdidas*. Barcelona: Anagrama.

Pinar, W. F. (2011). *What is curriculum theory?* (2nd edn). Oxford: Routledge.

Sosenski, S., and Jackson Albarrán, E. (Coord.) (2012). *Nuevas miradas a la historia de la infancia en América Latina: entre prácticas y representaciones*. México: UNAM.

Weldon, G. (2013). *Superman: The unauthorized biography*. Hoboken, N.J.: Wiley.

2

Conceptions of Childhood in the *Vecindad*

Victoria Parra-Moreno*

Introduction

Screenwriter, director, and actor Roberto Gómez Bolaños (Chespirito) cast the lives of characters on the margins of society—lawbreakers, underprivileged, the insane, unsuccessful superheroes, elderly, and children—at the center of his shows. The perception of such figures as minorities, positioned on the fringes of society, is molded by the prevailing organization and moral norms of the dominant culture. Thus, it is interesting that Chespirito's characters generated such enormous and enduring attraction from television audiences—the same mainstream culture that casts these characters to the margins—throughout Latin America. This is especially true of the reception to *El Chavo del Ocho*,[1] Chespirito's TV sitcom whose protagonists are a group of "children," portrayed by adult actors, who live in the *vecindad* (a word used in Mexico and other Latin American countries to designate low-income neighborhoods).

Produced from 1971–1980, *El Chavo del Ocho* is a sitcom that continues to be part of the programming grid of channels in different latitudes, watched not only by adults who were children in the 1970s, but by twenty-first-century children as well. Thus, the TV show is capable of transcending generations and localities despite its setting in a particular social, cultural, political, and economic period. As Kincheloe and Steinberg (2000) note, situating the examination of such cultural products and practices within the study of childhood education is productive, since these discursive practices contain knowledge, values, and principles about what is possible or not to say, who is allowed to tell or must listen, and whose experiences are valid or irrelevant.

* The co-editors would like to thank Sarah van den Berg for her assistance editing this chapter.

This analytic undertaking examines how *El Chavo del Ocho* participates in the discursive practice and production of childhood, exploring the notions of childhoods presented and constructed through the silver screen. It considers how characters such as El Chavo and his friends are produced in relation to dominant discourses about children, and how, in turn, they contribute to a public discourse and imagery that produces knowledge and beliefs about childhood, knowledge that is historically and politically situated. The ideas emerging from these reflections are neither intended to signal progressive or detrimental images of childhood, nor to argue for a particular moral perspective about childhood. Instead, this chapter stresses the social construction of childhood that emerges from interconnected images and discourses about children. Furthermore, as social practices, these images and discourses of children have implications for children's capabilities and possibilities in society. Thus, I argue the cultural production of *El Chavo* challenges linear and deterministic depictions of children.

To convey these reflections, I tinker with the structure of a screenplay for an episode of childhood in three scenes featuring the children from *El Chavo del Ocho* as its protagonists. Preceding the episode is an overture, which introduces the leading characters of the show. In Scene One, I foreground the adults behind and on the scenes, who bring to life the children of the *vecindad*. In Scene Two, I explore the relationships between images of childhood from *El Chavo* and a common, romanticized (Western) ideal of childhood. In Scene Three, I consider the implications of the fact that children are the primary audience for this cultural production that has transcended generational and regional borders in Latin America.

Title sequence

El Chavo and his friends are the daily companions of real children throughout Latin America, and some introductions to these familiar figures are in order before diving into the analysis of childhood. To assist in this introduction, I draw on my own memories as I revisited the show, as well as collective public descriptions of the characters via open source websites.[2] The mixture of sources and information underpins the idea that childhoods are evolving social constructs. These characters have been introduced to this narrative mainly through adult recollections, which are undoubtedly tinged with particular feelings and beliefs about childhood.

Chavo del 8

Chavo, also known as Chavito or El Chavo, is the protagonist of the sitcom. He is an eight-year-old orphan, who lives inside a barrel located in the humble courtyard of the *vecindad.* The show never reveals his real name; he is referred to instead as "Chavo," a common diminutive for young boys in Mexico. Several episodes also imply that he lives in Apartment 8, but this is never directly shown on screen. El Chavo is a caring, creative, naïve, and shy boy, who gets excited easily, especially when he is invited to play. He always volunteers to help his neighbors, but the adults and children in the *vecindad* exhibit widely differing attitudes towards him, from expressions of affection to physical and verbal abuse. Chavo constantly mentions how hungry he is, and often makes unintentional mistakes. He also suffers a rare (fictional) condition called *garrotera,* in which his body completely immobilizes when he gets scared until somebody throws a glass of water on his face.

Chilindrina

She is the only girl of the steady group of friends in the *vecindad* and the only daughter of Don Ramón, a widower who is known for not paying his rent on time. Chilindrina is a short, friendly, outspoken, and witty child. A clever girl, she is capable of manipulating adults and other kids to get what she wants. Sometimes she likes to make fun of her friends. Chilindrina participates eagerly in all games, including stereotypically male games. She lives in a household with scarce economic resources.

Quico

His name is Federico, but he does not like to be called by his real name. He is the only son of Doña Florinda, a single mother who is overly protective of and spoils him. His distinctive body traits include his fat cheeks and crooked legs, for which his friends tease him. In comparison to the other children, he has more economic resources, so he is seen with a variety of toys. He does not like to share with his friends, but is generally a nice and naïve boy. His mother always refers pejoratively to her neighbors as "chusma," meaning that they are people of bad manners or poorly educated, and Quico often parrots this expression.

Ñoño

He is the only child who does not live in the *vecindad,* but often visits because his father is the landlord. Ñoño attends the same school as the other children but constantly demonstrates that he is more knowledgeable and better mannered than others. Although he is often mocked for being overweight, he is still always eager to play with the rest of the children.

The set of characteristics gathered in these descriptions creates an image of the children's social standing and offers a glimpse into their internal worlds. Each of these characters has social or physical traits that deviate from the romanticized image of the child growing up in a two-parent family, sheltered from harsh conditions of hunger or abuse. By constructing characters that depart from social norms and expectations, *El Chavo del Ocho* provokes critical reflection on the knowledge exposed—and circulated—by these images.

Scene 1: Imagining childhoods from adulthood

Examining the cultural phenomenon of *El Chavo del Ocho* entails reflecting on the architects of these images of children. *El Chavo del Ocho* was created by Roberto Gómez Bolaños, a Mexican actor and comedian, who also starred in the show as El Chavo. It is also significant that the creator of the show is from Mexico and is a humorist. These two qualities contribute to the creation of a space in which adults could talk about children living in poverty, in a way that complicates the social ideals of childhood. Adults have not only written the child character parts, but also act in those roles. This is significant because adults inhabit social positions that provide them with the authority, power, and knowledge to talk about the child and childhood, which in turn produces normalized practices of childhood and distinctions and divisions between types of children. The prevailing notions of childhood have been constructed and disseminated through the dominant voices of adults, whose ideas should be acknowledged as situated in an arrangement of culturally shared ideologies about the nature of children, adults, social institutions, and society (James and Prout, 1990). As Ariès (1962) notes, the idea of childhood is a social construction that "corresponds to an awareness of the particular nature of childhood, that particular nature which distinguishes the child from the adult" (p. 128). As a social construction, the notion of childhood has evolved over time, which in turn allows us to examine the extent to which such conceptualizations reflect the relationship between adults and children in terms of power and knowledge of each other, a relationship that produces the child as the "Other" (Bhabha, 1996).

In the very act of creating a television show about and for children (even though it was initially intended for adults), the creator of *El Chavo* and the adults acting its child characters have made themselves vulnerable. The process of representing childhood(s) for the television screen inevitably draws on personal memories of the past, which reflect specific life experiences and are shaped by particular economical, social, and historical conditions. Therefore, what the screenwriter or actor conveys to today's audiences reveals fragmented images and reminiscences of his or her own childhoods. El Chavo and his friends, conceived of and embodied by former children, are characters that tell us about the adults and their memories as well as what they have forgotten, or omitted, in creating characters for the screen. As Gullestad (1996) suggests, "through the poetic imagination of the adult story, it may sometimes be possible to catch glimpses of what it is like to be a child (. . .) dialogue between the child the author once was and the adult he or she is now" (p. 19). Thus, the story of *El Chavo del Ocho* is also a story of adults, and of their childhoods.

Although the creator of *El Chavo* did not live in a *vecindad* as a child, nor was his family poor, his childhood, as well the childhoods of the cast, took place in a sociocultural milieu in which social differences and inequalities were present through cultural productions like *Cantinflas*' movies. In the late 1930s, Mexico witnessed the onset of the acting career of Mario Moreno, who gained international recognition for his character in the eponymous movie series *Cantinflas*. The character portrays a 'peladito' (peeled), which stands for a poor slum-dweller. *Cantinflas*, as *El Chavo del Ocho*, enjoyed an enthusiastic reception throughout Latin America, and it could be argued that it paved the way for *El Chavo del Ocho* by familiarizing television audiences with images of underserved *barrios* and their inhabitants. As Williams (2011) suggests, *Cantinflas* used mass media to make issues of social inclusion and exclusion palpable to a large audience. *El Chavo*'s creators grew up with *Cantinflas* on the screen, and following its example, *El Chavo del Ocho* could be seen as continuing the project of exposing poverty and social inequalities through characters that are explicitly positioned at different social statuses and life trajectories.

The representations of children in *El Chavo* are shaped not only by the memories of adults but also by the creative possibilities and transgressions of comedy. There has been extensive scholarly work in the past decades (see Freud, 1960; Zijderveld, 1968; Romanos, 2012) that analyzes the relationship between humor, social conditions, and discontent, considering how and why humor works as a medium for ideas, feelings, and needs related to acknowledging and acting upon social injustices. It follows, then, that *El Chavo del Ocho* is a comedy that reveals harsh situations suffered by children and families, positioning adults as doing social criticism.

The use of comedy makes it possible to represent social issues such as poverty and life on the margins of society, but also colors the reality of these conditions. The question for this reflection concerns the latent consequences of the creation and dissemination of particular types of childhood through humor. On the one hand, comedians portraying social issues, as in *Cantinflas* and *El Chavo*, have been able to open a public space in mass entertainment for showing social class relations and conflicts through central characters that materialize a history of social production and organization. On the other hand, showing marginalized childhoods through the genre of comedy creates a caricaturized rendering of reality. Opportunities to laugh both at and with the characters downplays the severity of the hardships they face and potentially spreads the perception that poverty is an unfortunate, but tolerable, condition.

Thus, the adults creating *El Chavo del Ocho* play dual, contradictory roles in both contesting and contributing to injustice. As critical social actors, they employ comedy to expose and challenge inequalities in society. Yet the 'making fun' of poverty and social injustices that the children face overshadows their needs and misrepresents social inequalities. For example, the show both caricaturizes and naturalizes the state of a constantly hungry orphan child living inside a barrel, which the adults around him do not treat as problematic. That the adult characters in the show do not take issue with Chavo's condition stresses the point that adults often define the life trajectories of children through their own power and decision-making. The ways in which adults represent children, especially through mass media, has similar power and consequences for the discursive practice and production of childhood. There is a fine line between employing humor to underscore social inequities, and exploiting humor to laugh about children and poverty. In this sense, portraying children who face hardships could create opportunities to engage in social critique, while these humorous images could simultaneously naturalize the existence of social inequalities and diminish the urgency of social problems.

In the next section I draw on scenes in the show to discuss the way in which ideas about childhood are constructed through multiple layers and multifaceted characters. The examination of childhood means exploring the ways in which children are imagined and depicted, the labels that emerge from these representations, and the possibilities these images open up for young people in society.

Scene 2: Childhoods in the *vecindad*

The hypothesized dichotomy between childhood and adulthood, as two distinctive stages of life in which people have different capabilities, interests, needs, motivations, preferences, responsibilities, and knowledge, can be helpful as a starting point from which to explore the socially constructed notion of childhoods. Furthermore, taking this dichotomy as a starting point for analysis could potentially complicate the binary of childhood/adulthood. Through this inquiry into the show, I conclude that it is precisely the characteristics of the marginal nature of childhoods portrayed in *El Chavo* that create the possibilities for interrogating idealized notions of childhood. Children living in the *vecindad* demonstrate social practices and individual characteristics that allow the audience to question the politics of childhood. In this sense, viewers interrogate themselves on how one should think, see, and act towards children, especially marginalized children, while also examining how adult authority and power is exercised.

According to Ariès (1962), the contemporary idea of childhood in the Western world arose hand-in-hand with modern notions of education, family, and progress. This prevailing concept of childhood builds upon a few distinct characteristics. Considerably influenced by Rousseau's (1762) work, the contemporary child is essentially characterized by the biological traits of the young human which emphasizes the immaturity and development of physical and mental abilities. The child is thus treated as a vulnerable being, which immediately situates children in a particular relationship with the adult world (Frankel, 2012). As Hockey and James (1993) reason, the child is set apart as the "Other," relative to the adult world as the norm or standard. As such, the child inhabits a social world that is spatially, temporally, and qualitatively different from that of adults. Implicit in this image of childhood as a separate realm is the idea that children are too inexperienced to handle the adult world, and need to be separated for their own protection. This belies the assumption that young people need to be sheltered from the difficulties inherent to adulthood, as well as shielded from harmful adults. By labeling children as incomplete beings, actions such as surveillance and regulation are deemed necessary for their own protection (Walkerdine, 1984). Furthermore, images of adults reify them as the gatekeepers of children's development and well-being. Underlying this construct of childhood are distinct perceptions, emotions, and ethical positions about what is right and wrong for children. These constructions run the risk of being taken as universal truths, controlling peoples' lives and limiting actions and expectations for children. The following analysis examines the assumptions underlying or constructed by characteristics of the children living in the *vecindad*: El Chavo, Chilindrina, Quico, and Ñoño.

Living in the *vecindad* del Chavo (and complying to social norms and expectations)

All children in the show exhibit some common attributes that are valued and expected of childhood, such as innocence, playfulness, and overpowering emotions in moments of joy and sadness. Chavo and his friends also conform to the social norms as they play and attend school. In addition, by portraying children of the *vecindad* in compliance with social norms for childhood, the show contributes to the naturalization of some characteristics as if they were innate to children, independent of their social class and culture. In this sense, children growing up in poverty appear to share characteristics with children living in different socioeconomic settings.

Furthermore, these images clearly distinguish between the child's and the adult's world, implying children's affects and activities—play, innocence, uncontrollable crying, and being students—are expected and accepted during this stage of life. In particular, the affective world of children characterized by quick emotions governing their behavior, is considered intrinsic to childhood. In addition, as Cannella (1997) argues, the discourse of innocence denotes a lack of knowledge, which is an adult trait. The show's portrayal of children as innocent could reinforce the notion that children need adult supervision. In other words, lack of knowledge is a deficiency that signals a difference with the adult world, as adults have the supposed maturity to handle stressful situations through acquired wisdom and more developed coping mechanisms. Such a portrayal of childhood as a state of innocence and lack of knowledge contributes to the imbalance of power between children and adults when the children are exposed to the imposition of adult control, surveillance, or punishment.

Yet, while *El Chavo del Ocho* contributes to the social image of childhood by showing children complying with the norms and expectations for this stage of life, it also portrays some adults engaging in childlike behavior. For example, the grown men Don Ramón and Profesor Jirafales are sometimes seen to be crying and Quico's mother is often seen as incapable of controlling her anger when she screams and slaps Don Ramón. The show thus defies the strict dichotomizing of childhood and adulthood and relativizes the very images of childhood to which it also conforms.

Living in the *vecindad* (and deviating from social norms and expectations)

El Chavo del Ocho also disrupts popular images of childhood as its young characters engage in behaviors that deviate from social norms and expectations for their age. Not only do the children of the show live in social conditions that challenge the way in which society typically imagines childhood as a stage protected from the evils of the world, they also behave in ways that society would not hold in high regard or associate with a romanticized ideal of childhood. A clear example of one of the deviant behaviors is mischievousness. Chilindrina is the character who most often seems to enjoy making fun of her friends, by mocking them or playing pranks. I deliberately use the term 'enjoy' here, as the character derives great pleasure from these actions. In several episodes, Chilindrina ridicules El Chavo by telling her friends that "he got late to the distribution of brain" to imply he did something dumb; or taking things away from her friends and running away while saying "*matanga dijo la changa*" (which roughly translates to "you snooze you lose"). These behaviors are far from a romanticized image of children as virtuous, "angelic beings." Chilindrina, on the contrary, shows a child who is entertained by her bad behavior. Despite the fact that other cultural productions also show children whose behaviors deviate from social norms (e.g. Bart Simpson, Dennis the Menace), I argue that in the case of *El Chavo*, the socially undesirable behavior of Chilindrina challenges not only the image of an ideal child but also the notion of girliness often promoted by cultural productions for children, in which girls are usually portrayed as more well-behaved, mature, and kinder than boys. In this sense, the naughtiness of Chilindrina's character has a double function in resisting pervasive morality about children and girls, making *El Chavo del Ocho* a disruptive cultural production regarding gender and childhood.

But the show not only challenges expectations of innocence in gender and childhood through misbehavior. The very fact that *El Chavo del Ocho* revolves around an orphan child surrounded by adults who do not take him into their direct care challenges the idea that adults are responsible for ensuring children's safety and well-being. The sitcom's plot breaks with the assumption that children are in need of adults' skills and resources to survive and thrive, since they supposedly lack the capability to care for themselves. El Chavo is depicted as very young independent person, who not only takes care of himself, in terms of meeting his basic needs, but he also chooses to attend school without the supervision or guidance of someone with a supposed moral superiority. The fact that El Chavo is constantly begging and betting to get food is not an issue of concern for the adults around him, further illustrating the incompetency of the adults in securing the welfare of the children around them.

Further challenging the ideal of childhood as a protected stage of life is the fact that children are often victims of physical and verbal violence. For example, Chilindrina's father Don Ramón is seen in each episode infringing on the children's rights. He teases Ñoño for his large shape, punches El Chavo after the child has done some mischief, and pinches Quico for his silly comments. To complicate matters, other adult characters who witness these abusive behaviors often ignore Don Ramón, particularly when the child is El Chavo. Quico is the only child who is defended by his mother from other adults. Yet, her defense always involves violent verbal and physical responses such as derogatory comments about the others' lack of resources and education. Not only is violence met with violence, but social labeling and discrimination are also layered onto incidents of child abuse.

These examples signal that children in the *vecindad* live in a social and emotional environment in which adult violence and neglect towards children is permissible from the people who are supposed to follow particular social mandates and have the competency to care for them. Consequentially, the show presents the children of the *vecindad* not as being less competent and powerless than adults, but rather capable of facing harsh conditions—including those adults who threaten their security. By creating images of neglected children surviving and even having fun, the show problematizes the binary construction of weak, inexperienced children and sensible, protective adults.

As *El Chavo del Ocho* tells the stories of mostly single-parent families living in the particular social context of the *vecindad*, the show contests romanticized ideals of childhood. A significant part of the children's daily activities happens in the absence of adults. This could be read not just as the children's desire for independence from the adult gaze, but also a lack of desire on behalf of the adults for parenthood.

El Chavo del Ocho also complicates notions of childhood through its portrayal of how social class materializes in shaping the opportunities and experiences of childhood. Throughout the episodes, the children—particularly El Chavo—constantly exhibit an awareness of social class. In an episode that takes place in the classroom, the teacher asks students what kind of animals eat everything, intending for the answer to be "omnivores." El Chavo answers "rich people," which implies that he sees wealth as a trait of those who have access to essential resources. In another scene, Ñoño, the son of the landlord, responds to a personal situation that negatively affected him with the expression "I will tell my father to raise your rent." His response implies that his participation in a higher social class enables him to act in ways that have power over other people.

Thus, *El Chavo del Ocho* constructs a complex image of childhood through struggles between differing knowledge and values about children, their abilities, and their relationship with adults. The show not only depicts children defying socially expected behaviors and norms, but also contests norms of parenting and filiation through its portrayal of child abandonment and neglect as accepted characteristics of human beings. Images of both children and adults in the show are far from romanticized fairytale images in which characters lived happily ever after. Instead, through a simple and repetitive plot, the sitcom introduces characters that epitomize the lives of marginalized children and adults whose life trajectories are rooted in particular settings, often characterized as underserved, poor, and vulnerable. The marginalized context in which childhood (as well as adulthood) materializes pervades these scenes. The context of the *vecindad* presents both opportunities and challenges for the lives of children and families; these are expressed in the material resources and social spaces that allow for children to be and become subjects in the absence of the adult gaze. The initial provocation of challenging the binary of childhood and adulthood shows the limitations of thinking of childhood merely in opposition to adulthood. Instead, adults and children in the popular sitcom transgress assumed borders between childhood and adulthood. As Kincheloe (2000) indicates, the show's producers and actors seem to be 'adulting' the child while 'infantilizing' the adult.

Scene 3: Children as the audience

In its heyday, *El Chavo del Ocho* reached millions of people throughout Latin America. Despite the fact that *El Chavo del 8* was not initially created for children (TV Azteca, 2012), for more than two decades the show continued to capture the attention of young audiences. Thus, the show is not only a cultural product that depicts the cultural practice of childhood, but a discursive practice that conveys images and discourses of childhood to children themselves and thereby participates in the lived constructs of childhood.

What makes these images and discourses of childhood so compelling to children and adults throughout Latin America? As previously discussed, I argue that *El Chavo del Ocho* contributes to complicating idealized notions of childhood by engaging spectators in common images and discourses that are historically, socially, culturally, and economically situated. In this sense, the sitcom provides opportunities to talk about issues such as social class, social differences and disparities, and the neglect and abandonment of children. I argue that the show also appeals to children by connecting with them on an emotional level and engaging their imaginations in compelling issues.

Children's affective realm

On the one hand, the enthusiastic reception of the series among children suggests that the show captures their imagination and desires. Some of the pleasure in the viewing experience comes from watching Chavo and his friends play, in ways, it could be argued, that mirror the social practice of play familiar to and enjoyed by children in different social and cultural contexts. However, the show also allows viewers to experience feelings of pleasure that arise from envy, selfishness, and bullying. Characters appear to enjoy both playing pranks on and mocking other characters. Thus, children who watch the show can identify with a double morality: that is, they can identify with socially valued elements as well as with "guilty pleasures." Presented with characters that are allowed to transition between suitable and inappropriate behaviors, young viewers are given an opportunity to vicariously experience with fantasies and desires of children who are not subjected to the moral norms of dominant society. An example of this deviant behavior is when *El Chavo's* characters mock and upset adults, either intentionally or accidentally. A common scene includes Chavo's insults and "inappropriate" comments such as calling his schoolteacher "Maestro Longaniza," referring to his tallness, which roughly translates as "Professor Long-sausage." In addition, the sitcom's characters often hit adults accidentally while playing. The possibility of enacting these actions on adults, presented in entertaining scenes, might appeal to children's fantasies.

On the other hand, *El Chavo del Ocho* presents children with a world in which they are abandoned, left to fend for themselves, and physically and verbally abused by adults. Instead of reflecting why the adults present these images to children, it is productive to reflect on the possible emotions these images of abuse and neglect might evoke. Children might feel sorrow and anger—emotions opposite to pleasure. There is a very popular episode in which El Chavo is expelled from the *vecindad* because he is suspected of theft. This appears unjust, since El

Chavo's rights have already been infringed upon by abuse and neglect, and triggers the audience's sympathy. However, even as young viewers experience unpleasant emotions, they are also provided with emotional support as the characters solve their transient problems and most of the time the show culminates in a happy ending. This could be thought of as an illusion of a future in which things always "work out," designed to make children believe that hardship is provisional.

Through the broad spectrum of affective experiences presented by *El Chavo del Ocho*, children not only enjoy their time in front of the screen but also experience varied emotions in a sequence that provides closure. Pleasure, grief, fear, and other emotions intermingle in ways that more closely resemble the affective world that children actually experience rather than a romanticized image of childhood where innocent children are sheltered from the realities of the world.

Introducing children to a complicated and implicated social life

El Chavo del Ocho traverses a wide range of social issues related to children, family, and community life, such as growing up in single-parent households, experiencing the death of family members, and facing poverty and social inequality. Through comedy, the adult actors and screenwriters present young viewers with a complicated world, in which children are emotionally and socially implicated. Children who have experienced some of these issues can connect to this imaginary world. The characters do not offer the audience Hollywood-like lives, "dreamed" places, royalist identities, and narratives of a flawless society. Instead the show depicts the everyday conditions and emotional experiences of slum residents' lives, which impacts children in different affective registers.

Commonalities between the show and the viewers' real lives could be why *El Chavo del Ocho* appeals to such a large audience: even if children might have not experienced poverty directly, it is likely they have had some experience of joy, fear, injustice, and friendship. As a cultural artifact, *El Chavo del Ocho* contributes to the making of the self, as well as the making of social relationships. The young viewers of the show are socially and culturally situated; they watch, read, and represent the images and discourses through their personal and social resources. The performance of actors further offers children experiences and opportunities to signify their own lives via elements that "talk" to children, as well as through experiences that invite children to "feel" childhood. The show provides children with constant representations of pleasure and pain; therefore, they are continuously "affected" by and implicated with El Chavo and his friends.

Final scene

El Chavo del Ocho, through stories of children and a common structure in which characters engage in a repetitive pattern of actions and events (e.g. Chavo being hit by Don Ramón, Chilindrina crying loudly, Quico's mother slapping Don Ramón after a misunderstanding), depicted cultural and material difference between social groups. A clear example is the series of episodes aired in 1979, in which the neighbors moved into the landlord's house when repairs were being carried out in the *vecindad*. In one of these episodes, El Chavo goes to Ñoño's bedroom and he is amazed by the large number of toys: "Where am I? Isn't this Santa's house? . . . I have never seen so many toys." He is then asked if he brought some luggage; he replies that he brought a few marbles and a slingshot, but no clothes as he is wearing all he has. This scene exemplifies how the producers of *El Chavo del Ocho* presented a social, cultural, and economic reality that, despite it being produced in México, is not exclusive to this geographical location. The plot introduces phenomena that are familiar to other countries in Latin America, which helps to explain why it was so compelling to audiences throughout the region. This is reinforced by the repetitive structure of the sitcom, as this feature strengthens the audience's familiarity with the show, contributing to the show's roots in the collective memory of generations who fondly remember El Chavo and his friends, and creating an emotional bond that transcends Latin American borders.

Some adult critics have argued that *El Chavo del Ocho* is a negative influence on children due to its frequent scenes of abuse, mockery, and violence (Burg, 2008; TV Azteca 2012). Despite these criticisms, the show and its "stickiness" quality in children's and adults' memories invites viewers to explore the ways in which childhood is imagined and articulated, and how contrasting philosophies and stances are merged in the portrayal of childhoods.

In centering his sitcom around childhood stories, it could be said that Chespirito puts the issue of childhood on the table "*sin querer queriendo*," or without meaning to (that could be translated as "accidentally on purpose", Chavo's popular saying). In its depiction of marginalized childhoods, the show represents the adults' perspective or construction of childhoods. Through the creation of archetypical characters, the show has been able to introduce its Latin American audience to the highly relevant and pervasive issues of social classes.

The child characters could also be representative of other social characters: Quico typifies the spoiled child; Chilindrina, the mischievous child; and Ñoño, the privileged child. While it uses the familiarity of characters to connect with audiences, the show also introduces a unique character—El Chavo. El Chavo's character is one that fairy tales omit, since issues of abandonment, neglect, and violence are never solved by the end of the show. By humorously presenting a child that challenges stereotypical characters of children's shows, *El Chavo del Ocho* opens up a space to discuss how social inequalities shape childhood.

Consequentially, *El Chavo del Ocho* complicates the notions, images, and discourses of childhood (and by extension the images of adulthood). This has political implications, as the assumed characteristics of children have often determined their social roles and rights (Purdy, 1992). Further, the tales of the children in *El Chavo del Ocho* express the stories of marginalized lives; it uses comedy to introduce and interrogate challenging topics. By representing marginal children, we—the audience—are offered the possibility both to examine the discourses and institutions of the child and childhood, and to reflect on the possibilities for children's actions and trajectories. This implicates the audience in the construction of childhood and the reality of children's lives.

Deleted scenes

This analysis of how adults construct childhood through *El Chavo del 8* has been conducted in the absence of the voices of children. Highlighting this absence signifies that children are special kinds of humans, separate from adults. Yet, it also emphasizes adults' authority to imagine, narrate, talk, and produce knowledge about children. This work has not tried to be a comprehensive review of constructed childhoods, but a perception of the complicated nature of the discourses of childhood, particularly when these intersect with issues of poverty and inequality.

Notes

1 Translated as "The Lad of (Apartment) 8."

2 Wikipedia, El Chavo del 8. Available at: https://es.wikipedia.org/wiki/El_Chavo_del_8#Cr.C3.ADtica. Chavo del 8. Homenaje al programa # 1 de la TV humorística. Available at: www.chavodel8.com/

References

Ariès, P. (1962). *Centuries of childhood: A social history of family life*. New York, NY: Knopf.

Bhabha, H. (1996). "The Other Question." In P. Mongia (Ed.). *Contemporary postcolonial theory: A reader* (pp. 37–54). London, UK: Arnold.

Burg, J. (2008). *La Violencia en 'El Chavo del Ocho.'* Munich, Germany: GRIN Verlag. Available at: www.grin.com/es/e-book/134076/la-violencia-en-el-chavo-del-ocho.

Cannella, G. S. (1997). *Deconstructing early childhood education: Social justice and revolution*, (Vol. 2). New York, NY: Peter Lang Publishing.

Frankel, S. (2012). *Children, morality and society*. Basingstoke,UK: Palgrave Macmillan.

Freud, S. (1960). "Jokes and their Relation to the Unconscious." In *The standard edition of the complete psychological works of Sigmund Freud, Volume VIII* (1905).

Gullestad, M. (Ed.). (1996). *Imagined childhoods: Self and society in autobiographical accounts*. Oslo: Scandinavian University Press.

Hockey, J., and James, A. (1993). *Growing up and growing old: Ageing and dependency in the life course*. London, UK: SAGE Publications Limited.

James, A., and Prout, A. (1990). *Constructing and reconstructing childhood: Contemporary issues in the sociological study of childhood*. London, UK: Falmer Press.

Kincheloe, J. L. (2000). "Sólo en casa y malo hasta la médula: surge una infancia post-moderna." In J. L. Kincheloe and S. Steinberg (Eds). *Cultura infantil y multinacionales* (pp. 45–64). Madrid, Mexico: Morata.

Kincheloe, J., and Steinberg, S. (2000). *Cultura infantil y multinacionales*. Madrid, Mexico: Morata.

Purdy, L. M. (1992). *In their best interest? The case against equal rights for children*. Ithaca, NY: Cornell University Press.

Romanos, E. (2012). "The Strategic Use of Humor in the Spanish Indignados/15M Movement." In *Politics and Protest Workshop CUNY Graduate Center, New York, USA* (Vol. 4).

TV Azteca. (24 November 2012). *La historia detrás del mito: La vecindad del chavo*. México: TV Azteca.

Walkerdine, V. (1984). "Developmental Psychology and the Child-centered Pedagogy: The Insertion of Piaget into Early Education". In J. Henriques, W. Holloway, C. Urwin, C. Venn, and V. Walkerdine (Eds). *Changing the subject: Psychology, social regulation and subjectivity* (pp. 153–202). London, UK: Methuen.

Williams, G. (2011). *The Mexican exception: Sovereignty, police, and democracy*. New York, NY: Palgrave Macmillan.

Zijderveld, A. C. (1968). "Jokes and their relation to social reality." *Social Research*, 286–311.

3

"Here comes Chavo! Everyone's watching the TV"[1] . . . Thinking about Difference and Alterity, Childhood, and Education*

Ana Paula Marques de Carvalho and Rita de Cássia Prazeres Frangella

It was on a Friday night at 6 p.m., on August 24, 1984, that the first episode of *Chavo*, entitled "Hunting Lizards," premiered on Channel 11, run by Brazil's Silvio Santos Group's old TV Studios.[2] The series—with characters like Chavo, Quico, Don Ramón, Doña Florinda, Chilindrina, Profesor Jirafales, Señor Barriga, Doña Clotilde, Ñoño, among others[3]—took on a life of its own among that programming network (now called SBT), with Brazil taking first place in terms of households watching the program among other countries, according to the show's actors (TV e Lazer, 2014). The opening quote of this article is from a popular newspaper that put out a special issue celebrating 30 years of the program being on the air in Brazil.

The story of a boy who lives in a *vecindad* and, as other boys do, plays around, plays pranks with his friends, goes to school, and interacts with other members of his community, came to represent our day-to-day lives, through a narrative that seems simple, common, and reflective of the daily lives of children in many parts of Brazil. Even though this boy was not Brazilian, his day-to-day life reflected many Brazilian childhoods and gained ground, becoming such a well-known phenomenon that some expressions and sayings of characters from the show have become part of everyday speech.[4]

Thirty-two years have passed and the current millennial Brazilian generation also laughs and cries with Chavo. "He's clumsy," noted the five-year-old son of one of the authors, who still watches the show with his father, and who (like viewers of all ages) still enjoys the shenanigans of this character. The series "Chavo" traversed the

* Translated from Portuguese by Rolf Straubhaar.

world and impacted Brazilians at the post-dictatorship moment in which Brazilian society was reconstructing its democracy and sought to reclaim a new generation's rights. In the 1980s, Brazilian television saw the opening of new pathways in its programming—previously censored *telenovelas* returned, the famed cartoons of William Hanna and Joseph Barbera were aired, and new 1980s programming (such as *Thundercats* and *He Man*) came quickly thereafter (Gonçalves, 2010).

We have appropriated the Brazilian version of the introductory music for the Chavo show in order to explain the task we have before us in this piece: "Here comes Chavo, Chavo, Chavo . . . Everyone's watching the TV." As we watch, what do we see?

Our focus in this reflection-based piece is not to explore what made Chavo such a TV phenomenon (not just in Brazil, but throughout Latin America, as can be seen through its audience numbers and merchandising throughout the region[5]), but to think alongside Chavo and the children the program focuses on. What are the central themes of childhood discussed in the program? What implications do these have for the study of childhood and education?

To explore these questions, we enter into dialogue with the larger literature on difference, culture, and alterity, particularly drawing on the work of Homi Bhabha, Arjun Appadurai, and Jacques Derrida. We bring to this debate themes of childhood highlighted within the Chavo program, and we argue that it is necessary to discuss childhood when experienced from a condition of alterity, to think of childhood as a liminal space of cultural production separated by difference. When we speak of difference, we do not only refer to age; we distance ourselves from a homogenizing perspective buoyed by the logic that comes from looking at childhood from an adult's perspective where childhood is seen as a state of minority or incompleteness that negates childhood as alterity.

At the same time, we find it important to highlight early in this writing a point which we propose as being an important characteristic of the show: we do not see Chavo as solely children's programming, directed only at children. Rather, Chavo has shown to be popular among adults as well as children, and cannot be shoehorned into a niche category of programming targeting a single specific audience.

At the same time, we are discussing adults playing child characters, an imitation that, in Bhabha's (2003) words, entails an attempt to represent childhood. There is a certain subversiveness in this representation in which an adult, acting as a child, brings a perspective of otherness, representing difference as it exists in the imagination and identity of a child. In this way, this is a sort of double representation, in which the act of imitating and interpreting childhood to a certain degree breaks with the opacity of the fantasy of childhood.

As a result, the characters in Chavo compel us to explore the themes of childhood that are processed through these differences and negotiations. In our understanding, these differences are moving in the relationship with the Other—in the alterity process—and involve cultural negotiations (Appadurai, 2004; Bhabha, 2003). The images, texts, and sensations mediated through our modern means of communication (here we are restricting ourselves to discussions of television) are part of this process of negotiation, in which subjects onscreen engage with the conduct of those people involved with and transformed by what they see through electronic communication. One point that must be understood is that means of communication (like television) have a large role in our dynamic production of meaning in the world, and thus it is important, when we think about childhood and education, to have a complete and complex discussion in which the media are recognized for their curricular role in that production of meaning (Appadurai, 2004; Pinar, 2011).

Chavo in cultural flows: Circulating images of childhood

The success and spread of this program in Brazil, and throughout the world, allows for a space–time perspective in which the images of childhood that we encounter in our day-to-day lives cross continually with one another, coming and going through past, present, and future. Appadurai (2004) understands this trend as a disjunctive transit of themes or meanings that are not restricted by the same temporal boundaries with which we are accustomed. These themes are in constant movement, and the means of communication through which they move (among them, television) act upon our imaginations, breaking with the traditionally constitutive features of our human subjectivity. In this framework one can see the continual construction and deconstruction of imagined "childhoods" and imagined children's worlds.

We see in Chavo interrogations, subversions, and transformations of the everyday, creating resources that make possible the construction of meanings of childhood in all types of societies. This is because Chavo allows for plots involving many types of possible lives thanks to the seductiveness of its characters, without losing plausibility. The means of television allow these constructed meanings to travel quickly into our everyday routines, acting as a tool through which every individual can imagine himself/herself as an ongoing social project. Through this lens, we understand that the concept of childhood, as outlined in the storylines of the program's characters, is a large part of what the show contributes to current dynamic cultural flows. Our understandings of childhood move within these flows, in a world of differences and contrasts.

In periods of accelerated globalization, Appadurai (2004) mentions that electronic communication gives a new texture of this world of differences and contrasts (which Appadurai understands as cultural), in which imaginative work turns every day into incessant flows shaped by the movement of *-scapes*. Appadurai (2004) uses the term *-scape* to point out the fluid, irregular nature of the meanings we produce in our imagined worlds. That is, our worlds are built through the construction and contestation of multiple universes, or *-scapes*, of meaning:

> I propose that an elementary framework for exploring such disjunctures is to look at the relationship among five dimensions of global cultural flows that can be termed (a) *ethnoscapes*, (b) *mediascapes*, (c) *technoscapes*, (d) *financescapes* and (e) *ideoscapes*. The suffix *-scape* allows us to point to the fluid, irregular shapes of these landscapes, shapes that characterize international capital as deeply as they do international clothing styles. These terms with the common suffix *-scape* also indicate that these are not objectively given relations that look the same from every angle of vision but, rather, that they are deeply perspectival constructs, influenced by the historical, linguistic, and political situatedness of different sorts of actors (Appadurai, 2004, p. 50).

The notion of the *mediascape* is a key concept to this discussion, as it helps us to understand the influences on our understandings of childhood. Appadurai (2004) mentions that *mediascapes* provide vast and complex repertoires of images that hold a tenuous line between fiction and reality, which contribute to narratives of alterity and other possible lives that can create the desire to acquire and move towards such "other" possibilities. The images of children created in *Chavo* contribute to the construction of imagined worlds of childhood, often fantastic ones, which can dislocate existing conceptions by allowing imagination to be seen as a social practice through which childhood can be seen differently.

With the support of Appadurai, we emphasize the importance of the notion of the *mediascape* in our analysis of the production of meaning which the Chavo program puts forth. After all, *mediascapes* contribute a focus on meanings created by images, offered to TV viewers, who experience these images as a series of elements (characters, plots, textual forms) that create a narrative reality upon the basis of which imagined lives can be formed (Appadurai, 2004). In *Chavo* we have an imagined world built on the flow of these *mediascapes,* which narrate the Other. However, in this flow, the imagined lives that are created vary from context to context, fracturing any attempt to fix meaning to these images and creating paradoxes in terms of what those images mean. This is because these *-scapes* (and to be clear we are not only including *mediascapes* here, but all of the *-scapes* that Appadurai (2004) notes, as they are not segmented, but work and interact together constantly) are continuously interacting fluidly across multiple dimensions, brought about by disjunctures and consisting of the incessant reinterpretation of our narratives. This constant flow of imagination contributes to the deconstruction of our (allegedly stable) image-based reference points. Our certainties are frustrated by the fluidity of transnational communication, which makes our relations one with another more volatile and more polarized, immersed in intense negotiations.

Thus in order to discuss the meanings of contemporary childhood, we have to go forward with the perspective that there are no more isomorphisms in the meaning of childhood and that we must free ourselves from limiting and primordial preconceptions of children. Of course, Chavo contributes to the construction of imagined worlds of childhood, worlds which are often fantastic but are always in displacement of concepts, ideas and conceptions that help us see childhood in different ways. Therefore, to think of childhood as existing in a world of disjunctive global flows is to see childhood arising from such flows and uncertainties; reflecting the image of chaos and knocking down the old images of order and stability which surround children.

It is interesting to reflect on the very contradictions of the characters used in the show: the orphan that is not so naïve or incapable; the girl who cries all the time, but brings with her cleverness and is always able to trick her friends; and the boy who always gloats to the orphan, but then cries, plays tricks, and hides behind his mother. In the midst of these flows, these movements of constant rupture, there is a production of differentiated imagetic perspectives, circulating images of childhood which are constantly negotiated according to their differences.

Acting out childhood: *Almost the same but not quite*[6]

What images are produced in our encounters with *Chavo* for us to see? The first, immediate response is a gesture towards children, childhood, and ways of being a child. With Bhabha (2003) we analyze the production of this image as a demand of identification, of being unto the Other that which, to the author, is always an image of identity—not a presupposed identity, but a demerged projection between presence and absence, representation and repetition, in a set of meanings that brings images produced as symbols, a vestige of childhood written upon the folds produced by these images.

Thus, as we engage with these images, the question of identity and alterity is rearticulated: the presentist logic attached to identity and image is shaken up and, as such, in the folds and gaps of time and space it becomes possible to break with identity as a self-fulfilling prophecy. These images of childhood register in a significant ambivalence that destabilize a polarized comprehension which models infancy in opposition to adulthood—that is, what an adult is, a child is not, in a demarcation according to what is missing, of not having become yet the *same*.

In this sense, Chavo, in his staging of childhood, erects himself as a symbol of childhood and, reading through a supplementary logic (Derrida, 2011), creates an articulation between presence and absence in its duplicity—his performance does not deal with presence or absence, but the ambivalence between a presence that is not absolute fullness and an absence that is not total absence, which in the space between these two emerges as difference:

> But the supplement supplements. It adds only to replace. It intervenes or insinuates itself *in-the-place-of*; if it fills to the brim, it is as if one fills a void. If it represents and makes an image, it is by the anterior default of a presence. Supplementing and vicarious, the supplement is an adjunct, a subaltern instance which *takes-(the)-place*. As substitute, it does not simply add itself to the positivity of a presence, it produces no relief, its place is assigned in the structure by the mark of an emptiness. Somewhere, something cannot fill itself up *by itself*, if not by allowing itself to be filled through sign and proxy. The sign is always the supplement of the thing itself (Derrida, 2011, p. 178).

There is the danger of the supplement, to which Derrida calls attention: the image-symbol that supplements, that occupies absence, that exceeds it. More comes, adding to the set of meanings, making it constantly move between paths, given the radical alterity of the Other which, unable to be fully captured, does not allow itself to be limited or fixed, frustrating a desire for truth. Supplemental logic allows us to infer about the question of representation of the image of the child, a gap of meaning that is yet to be filled. Through the representation of children in the characters of *Chavo*, we have a good example for thinking through the concept of childhood that is constantly moved by increases or modifications in meanings. Representation is the absent presence of meaning, introducing new meanings. Such representation is never complete. In our larger set of images and representations, the media (in this case the Chavo program) contributes to the increase in meanings of the concept of childhood, driving new symbols and, as a result, pushing forward new discursive formations that try to fasten meanings. These fastening attempts interrupt processes of meaning, producing ambivalences.

We will here bring in another perspective that will help us think through the set of meanings incited by *Chavo*—that is, reading *Chavo* through the construct of mimicry theorized by Bhabha (2003), mimicry which leads to the construction of an image. Here we address this—the production of images of childhood—and turn to mimicry to address another element which calls our attention, that of adults acting out childhood, adults acting as if they were children, with only a partial or always postponed presence of childhood in the resultant imitation.

Bhabha (2003) calls attention to mimicry as a strategy for taking hold of the Other, articulated through the production of difference—*almost the same, but not exactly*—that is at the same time rejected. It is this double articulation that produces an uncertain, partial presence. Of particular interest to us here is the alignment of Bhabha between mimicry and fetish, in a problematic ambivalence established by the anxiety of imagination and the normalization of this disturbance in the fetish object.

The idea of representability—recognition—establishes itself on the basis of rigidity. That is, we are recognizing something already given, an image of previous, fixed content. Thus representation/recognition is an expression of an apprehensible and visible reality. When reading this image as fetish, Bhabha leads us to think about how an image operates in its ambivalence: it ends in itself any recognition of difference while at the same time refusing such difference; it oscillates between domination and pleasure, anxiety and defense, articulating itself on the basis of a double image that threatens our desire for representation of an authentic, immutable reality.

It is "or," the mark of ambivalence, that threatens any strategy of fixation on the colonial subject. The "or" encircles and disturbs this representation, which allows us to conclude that this is not the expression of a mimetic practice, but of a discursive process of meaning-making. In Bhabha's words, "mimicry repeats more than it represents" (p. 132), which is also implied in "spaces of alteration and iteration (again, doesn't *iterum* come from the sanscrit *itara*, or 'other?'); from repetition of reproduction, or also, in space as a possibility of iteration and the exit of life to outside of itself" (Derrida, 2011, p. 255).

It is in this space of double articulation, where what Bhabha calls schism resides, which incites a game of meanings in which difference emerges as a radicality which cannot be camouflaged.

> The colonial presence is always ambivalent, split between its appearance as original and authoritative and its articulation as repetition and difference. It is a disjunction produced within the act of enunciation as a specifically colonial articulation of those two disproportionate sites of colonial discourse and power: the colonial scene as the invention of his toricity, mastery, mimesis or as the "other scene" of Enstellung, displacement, fantasy, psychic defense, and an "open" textuality. Such a display of difference produces a mode of authority that is agonistic (rather than antagonistic) . . . To recognize the *différance* of the colonial presence is to realize that the colonial text occupies that space of double inscription, hallowed—no, hollowed—by Jacques Derrida (Bhabha, 2013, p. 179).

The quote above highlights a few important points for discussion about the meanings of childhood driven forward by *Chavo's* characters. A boy living in a barrel–Chavo lives in a barrel–with wide pants held by loose and improvised suspenders walking around a *vecindad* with a wronged expression, probably projects the image of childish fragility, a boy overloaded by his socioeconomic surroundings, depending on adults to process his emotions and achievements. Pity, injustice, and fragility would likely be emotions that might overpower us at such a scene. However, if we consider the image described above as an open text, we can understand it and discuss it as a double text, which won't have the same meaning in each of the contexts in which it presents itself. Each identity is insufficient in terms of its Others. We are made up of the constant search to fill what is lacking. In this sense, the image of this boy in a barrel is always a symbol of movement which will never be the same. Symbols *in actu*, symbols of representation, unable to be determined or measured, rich in their acts of performative creation, in their contingencies, their fluidity and their production of differences in the battle for meaning.

The reproduction of childhood in these characters, in the mimicry of these actors, cleaves meanings, opens fissures, agitates differences and contrasts which are being explored in the mobilization of discourses, producing continuous effects on our understanding of the child and childhood.

In this sense, we are also able to reflect on the images and meanings of childhood that can be observed in the dialogues that occur between Chavo and his friends. In these dialogues, we are able to see the problematizing tension that we have been highlighting: the ambivalence and schism of thinking about childhood in its alterity.

This staged mimicry plays out like an ironic arrangement to disturb our normalized image of childhood. What we see on our screens are adults who, in the place of children, talking and looking like children, threaten the authority governing childhood by distancing themselves from a subaltern condition. It is children who are in the scene; in *Chavo*, it is children who talk with and about the world, who break a linearized and normalized logic that assigns the separate place and role of adult and child. Here the relationship between children and adults is shown in a dialogue that destabilizes a tiering of childhood distinguished by its lesser status to that of adulthood, but instead emerges as something else, as Other, in its difference. These are children who negotiate contingently, who subvert their condition of minority or subalternity in the fissures of these negotiations.

Don Ramón: How old are you, Chavo?
Chavo: Eight, why?
Don Ramón: It's that I don't understand how in such a short time you can become so stupid!
Chavo: Did it take longer for you?

Source: http://exame.abril.com.br/estilo-de-vida/noticias/10-frases-e-dialogos-iconicos-de-chaves 28/11/2014. Accessed July 2016.

To say this as subjects, these characters distance themselves from a simulated childhood drawn from a generalizing perspective, said by another and not by a child itself. Docility, a sense of coming into one's own, ingenuous smart-aleckness, and fragility are all contested in the exposition of transgressions of these generalizing ideas and from there, drawing on Bhabha (2003), it is possible to think that, in the space between mimicry and mockery, these characters break with the normalizing and unifying discourse of childhood.

Quico: Can't you see that I'm convalescing?
Chavo: Convale-what?
Chilindrina: Chavo, when Quico says he's convalescing, he means he isn't well, not really, silly!
Chavo: Ah, so when he's really silly he can come out to play? If that's all, he could have come out a long time ago!

Source: http://exame.abril.com.br/estilo-de-vida/noticias/10-frases-e-dialogos-iconicos-de-chaves 28/11/2014. Accessed July 2016.

We have to think contextually about what the Chavo program represents in terms of another look at childhood: bearing in mind that it was produced in the 1970s, focused on a child that is not separated from his social condition, unlike those of high status who are taken care of and preserved, but instead is placed (in a dialogic and relational sense), is produced and produces in a social arena marked by antagonisms, heterogeneities, and differences, breaking with a linear and apolitical narrative of childhood. To the contrary, this program enters the political arena as a form of alterity that produces hybrid, ever-moving forms that negate essentialized identities.

In this sense, one can note that the production of identity of Chavo is part of a process of alterity, through a practice of supplementing that breaks out of the larger homogeneous narrative of childhood and shakes the binaries that support it, going beyond that narrative and producing displacements—which, as Bhabha (2003) notes, are a hybrid production—destabilizing authority by making strange what is at the root of that authority: our rules of recognition, which are established through the enunciation of difference. Or rather, we put childhood forward as an expository question, no longer an object to be seen, observed, and appropriated. This occurs as we recognize childhood as other, as a form of alterity.

"Yes, yes, that . . ." [Eso, eso, eso. . .] or not this or that: Unpredictable thinking about childhood, education, and curriculum

Our reflections which dialogue with *Chavo* about the meanings of childhood act as interpolations to think about the relationship between childhood and education, and more particularly, about the production of curricular politics targeting childhood, which are a focus of the educational research studies the authors have previously conducted. This subject is very pertinent to the current moment we are living in Brazil, also present in other Latin American countries: a moment marked by a logic of curricular centralization, developed in defense of a perspective that pretends to be democratic, defending a pedagogical practice based in a national common standard that conforms to a vision of equality as oneness[7]—which we have problematized, on the basis of previous discussions on curriculum, culture, and difference. When what is common becomes singular, it wipes out any trace of difference.

Miller (2014) invites us to think with an analysis of our context, based in a culture of accountability and centralized evaluation. We align ourselves with her arguments regarding the need for curricular theorizings which expand our understanding of curriculum beyond standardized forms. Such standardized forms traditionally restrict curricular discussions to classification and measurement, keeping us from discussing other complex and relevant aspects of curricular content which help promote our understanding of education. In this sense, *Chavo* and the images or meanings driven by it can thicken our discussion of the following question: what kinds of education and curriculum work for and with childhood, observing it in its alterity?

To this question we add others: Who is the boy Chavo? Is there a single response to that question? How many different Chavos, Chilindrinas, and Quicos produce meanings in their day-to-day and are themselves produced in these processes? As such, do we want to find answers to these questions? Are there answers? Are there possible shared answers? In general, we would say that in our day-to-day, all of us involved in education seek ways to understand children, conjecturing a finished and ready model of childhood. We stipulate identities and meanings, often under the guise of sentiments like "this generation is not like others before it, that of my grandparents or my parents . . .". And truly, they are not the same. They never will be. Past, present, and future mix and the production of differences is constant. We are different, beings in process, unfinished, producing our own meanings about ourselves and the Other, us and the world.

Is it possible to subsume education to teaching, or curriculum to the formation of knowledges? We understand the answer to be no. Is it possible to understand children using fixed meanings, without considering them in the process of unfinished subjectivity in which subjects live and move? We say no.

These questions bring an important problematization in our present day, which relates to our purpose in this text, that of discussing relative meanings of childhood and their implications for education. In our previous research, we have sought a meaning of curriculum from a post-structural perspective that helps us discuss questions related to education without reducing it to the teaching of aprioristically fixed content, defended under the mantle of scientism. We understand curricula to be a political and discursive product, a cultural manifestation.

> In other words, that implies thinking the instituting movements of a curriculum do not take place based on a priori senses—the curriculum is shaped in the field of undecidability, in the articulation of different contexts. As a political-discursive production, I defend that curriculum policies are start with the negotiations and contests in which senses and interests are hybridized in curricular development, showing its state of unfinished political process. It is in the very act of negotiating that signifiers are built through dialog, many times based on conflicting consensus, as the senses are unstable, fluid and permeated by distinct demands and interests (Frangella, 2015, p. 27).

We opt for an understanding of curriculum based in an anti-realist perspective, or rather, we do not believe that there exists a reality that is delivered to us ready-made; rather, reality is constituted contextually and contingently produced through acts of power and interactions of meaning. In this perspective, curriculum is a meaning-based practice and must be read. And when it is read, it is betrayed, changed by social processes with disputed and negotiated meanings subject to power relations and, while it is being read and disputed, one part weakens and dies. In being translated, it loses part of its power. But its final meaning is always postponed.

In this way, we separate ourselves from perspectives that see curriculum as being attached to the notion of being an object, with pre-set definitions, linked to objective definitions or fixed knowledge to be passed on. We see our subjectivity as being created through the process, without an *a priori* that constitutes our being. There is no fixed or ready-made identity. We are constantly mid-scene in a moving performance, and we will never be the same in our different relations with the Other.

In this framework, we question the inherent essentialisms attached to the concept of childhood, which attach to early childhood education the label of fragile, incomplete, coming into being. The acts of power to which we referred earlier are not necessarily negative or coercive. We all produce power, even children. There is no center to this power, no structure which defines it or fixed positions for its subjects. The issue that motivates our work here is thinking through childhood in its condition of alterity:

> As soon as I desire I ask to be considered. I am for somewhere else and for something else. I demand that notice be taken of my *negating activity* insofar as I pursue something other than life, insofar as I do battle for the creation of a human world—that is, of a world of reciprocal recognitions. I should constantly remind myself that the real *leap* consists in introducing invention into existence. In the world in which I travel, I am endlessly creating myself. It is only by going beyond the historical, instrumental hypothesis that I will initiate the cycle of my freedom (Fanon, 1986 in Bhabha, 2013b, p. 30).

In our analysis of this citation, we follow Bhabha (2013) in asserting that the cultural dislocation of children and the social discrimination Fanon refers to are based in the desire for recognition of the cultural presence of children through a negating activity. Here we focus on the creative invention within existence and the staging of identity as iteration, the recreation of I, the reinvention that occurs in the battle for meaning within this identity.

Children dispute space, dispute meanings, and invent the conditions of their existence. They enact their identity as iteration, recreating themselves in the same process of subjectivation in which adults also participate. Children fight for the creation of a human world. They battle for reciprocal recognition. To paraphrase

Spivak (2010), who gave one of her works the title "Can the subaltern speak?" we ask: Can the child speak? We reflect on the concept of subalternity proposed by Spivak, bringing it into our discussion of the relations between childhood and education. In aligning childhood and subalternity we must remain aware of the polysemy of the term subaltern, which cannot be understood in a homogeneous sense. We understand and use the term following the meaning ascribed by Spivak (2010): that the dialogic nature of subaltern speech is obliterated. The subaltern is represented by another who speaks in their name, not because it is incapable of speaking, but because they are forbidden from doing so. Spivak argues that no one should speak for the subaltern who are in this condition maintained as an object, but that spaces of enunciation should be opened so that the subaltern can speak for and represent themselves. We understand that in observing a relationship based in alterity, the subaltern only encounters the legitimate Other while negotiating with them, which leads us to affirm (as does Pinar (2009)) that the content of education is alterity.

For this reason, it is important to think of curriculum as a complicated conversation, as a *currere* (Pinar, 2011). In an interview with Sussekind (2014), Pinar observes that curriculum is a conversation because in its use people are talking to one another, though this talk is complicated by the lack of transparency or autotransparency.

> It's complicated how opaque teachers and students are to themselves and to each other. Especially in a classroom with a certain number of students, it's like having a flash, you know: You don't think that the teacher tries to see everything, that he tries to scan, hear the silences and read between the lines: That the teacher, in looking side eye to the students, tries to see who is who: Well, this happens in a way that complicates the conversation. For example, if you are open to the reality of the other, you say things a little differently, without betraying the principle of what you meant to say; because of this, it is inevitably a complicated conversation. This conversation is also complicated by being informed, of course, by that which happened and happens outside of the classroom, like with students' families. This conversation is complicated because it happens between all of us in society (Süssekind, 2014, p. 31).

To immerse yourself in this complicated conversation requires an opening in which you can think of the child as Other—not on whom you can act, but with whom you can talk, argue, and make meaning. To the schools remains the challenge of recognizing the alterity of children, valorizing and encouraging their differences, engaging in lessons with them as subjects whose speech, writing, and forms of expression are their own. This conversation is an inner work that must come from within, in which the subjective experiences of children are woven contingently and marked by dialogicity, continuing enmeshed with each other and marked by differences, but no longer obliterated or negated.

Here we return to Chavo to think about this complicated conversation between adults and children. In fact, in the confusions of day-to-day life, Chavo's *vecindad* shows us this complicated conversation in which both adults and children are invited to encounter each other in the world in which they live as different beings. Children in this setting participate in this conversation, talking *to* education and not letting education speak for them (Pinar, 2011).

"Yes, yes, that . . ." [Eso, eso, eso. . .]—Chavo's catchphrase—we also do not see in binary terms, neither do we return with a parallel binary phrase such as "no, no, this." Going beyond binarisms, which mark as fixed the place of the child and the adult in educational processes, we seek to involve the reader in this complicated conversation with children, moving into the displaced frontier, the hybrid inscription of a contingent and interstitial between-place. Chavo invites us to do this: denaturalize your previous images of childhood and engage the production of meanings that problematize the "essentialization" of these images. In doing this, from there follows our defense/bet in a complex curricular texture, as an articulation and negotiation in which we move and act, producing ourselves as hybrids, on the unstable and contingent terrain in which subjects dispute the meaning of themselves.

Notes

1 This quote comes from the chorus of the theme music that would play on the Brazilian edition of *Chavo*.

2 Before gaining the rights to the four broadcasting networks that would make up the Brazilian Television System (SBT), the Silvio Santos Group had already owned Channel 11 in Rio de Janeiro since 1976, known as TV Studios (TVS). This channel was an important step to creating SBT, which went on air in São Paulo and throughout the rest of Brazil on August 19, 1981.

3 [Translator's note] Chavo, Quico, Don Ramón, Doña Florinda, Chilindrina, Profesor Jirafales, Señor Barriga, Doña Clotilde, and Ñoño, were called in Portuguese Chaves, Quico, Seu Madruga, Dona Florinda, Chiquinha, Profesor Girafales, Seu Barriga, Dona Clotlilde and Nhonho, respectively.

4 "Yes, yes, that," "I didn't want to want it," "No-one has any patience for me," and "OK, don't get mad" are a few of Chavo's most memorable phrases that are used in

day-to-day conversation in ways that indicate the spread and general appropriation of the program by the Brazilian public.

5 When we speak of merchandising here we refer to toys, school supplies, comic books, and other products featuring Chavo and other characters from the program, whose very ubiquity reinforces the argument that the program was immensely popular and successful. However, this is not the focus of this chapter. Nonetheless, the popularity of the show is evident in the number of websites and discussion boards dedicated to it, as well as the numerous books which have focused on the program as a central theme.

6 We borrow this expression from Bhabha, *O local da cultura*. Belo Horizonte: Editora UFMG, 2003, p. 134.

7 Here we refer to the Ministry of Education's process of producing and spreading a National Common Core curriculum for basic education in Brazil.

References

Appadurai, A. (2004). Dimensões culturais da globalização: A modernidade sem peias. Lisbon: Teorema.

Bhabha, H. (2003). *O local da cultura*. Belo Horizonte: UFMG.

Derrida, J. (1995). *A escritura e a diferença*. São Paulo: Perspectiva.

Derrida, J. (2011). *Gramatologia* (2nd edn). São Paulo: Perspectiva, 2011.

Frangella, R.C.P. (2015). Meetings, Dialogs and Interconnections in a Theoretical-analytical Perspective Design. Transnational Curriculum Inquiry, *v.*12, n. 2, pp. 24–30: Ottawa Curriculum Conference.

Gonçalves, M. (2010). Eu, a televisão e os anos 80. Anos 80: liberdade de expressão. [blog]. Available at: www.anosoitenta.com.br/televisao.html. Last accessed: Jul. 1, 2016.

Miller, J. (2014). Teorização do currículo como antídoto contra/na cultura da testagem. Revista e-Curriculum, São Paulo, v. 12, n. 03, Oct/Dec pp. 2043–2063.

Pinar, W. (2009) "Multiculturalismo malicioso," Currículo sem Fronteiras, v. 9, n. 2, pp. 149–168, Jul/Dec 2009. Available at: www.curriculosemfronteiras.org.

Pinar, W. (2011). *What is curriculum theory?* New York: Routledge.

Revista Exame. 10 diálogos icônicos de Chaves, 28/11/2014. Available at: http://exame.abril.com.br/estilo-de-vida/noticias/10-frases-e-dialogos-iconicos-de-chaves. Last accessed: July, 2016.

Spivak, G. C. (2010). *Pode o subalterno falar?* Minas Gerais: Editora UFMG.

Süssekind, M. L. (2014). *Quem é William F. Pinar*? Rio de Janeiro: De Petrus et Alii Editora Ltda.

TV e Lazer (2016). Atores de Chaves celebram os 30 anos de sucesso do seriado no Brasil. [site]. Available at: http://extra.globo.com/tv-e-lazer/atores-de-chaves-celebram-os–30-anos-de-sucesso-do-seriado-no-brasil–13684405.html#ixzz4Dq2naBKR. Last accessed: July, 2016.

Plateau II

El Chavo's Encounters with Other Latin American Societies and Cultural Artifacts

4

From the Picaresque Novel to *El Chavo del 8*

Carlos Aguasaco

The picaresque novel depicts the development and consolidation of urban capitalist societies in Europe and Latin America. Narrated in the first person and centered on a new type of emergent subjectivity, as correlated to new relations of production, it created a new textual typology similar to what we now call autobiography. Concepts such as the right to private property, the right to form and express one's own opinion, and the right to pursue social mobility, all of which seem natural or normal to Western societies were novelties to many people in the 1500s. Since its first appearance in the sixteenth century with *El Lazarillo de Tormes* (1554), the picaresque novel became the prime representation of a changing world that had stepped away from feudalistic relations of production and moved towards new capitalist relations of production and consumption with evident ideological and structural consequences such as social mobility based on the accumulation of capital and the eventual establishment of democratic forms of representative government. The free subjects who were entitled to sell their labor force in the market economy and invest the product of their work in pursuit of profit discovered that they also had the right to their own coming-of-age narratives, their own textual typology: the *bildungsroman*. To what should we attribute the popularity of the picaresque novel? In general, we can say that its wide reception and extended life was due to the convergence of three factors: the consolidation of an urban public demanding forms of entertainment related to its immediate reality and conditions of existence; the invention and widespread use of the printing press, which had caused an information revolution in the fifteenth century comparable only to the arrival of the Internet in the final decade of the twentieth century; the new proto-capitalist relations of production that transformed social structures through new ideas of social mobility based on individual effort and success. In *The Creation of the Media: Political Origins of Modern Communications,* Paul Starr (2004) calls the printed book "the original laptop" (p. 26), and goes on to explain how the dissemination of printed material created a "de facto public sphere" that included forms of communication developed "outside of the ruling elite" (p. 46). As a consequence of these changes, the picaresque novel emerged as a sort of open-ended epic, an epic of the individual subject in which the voice and the particular perspective of the protagonist displaces the discourse attributed to deities or nations in the preceding forms of epic narratives. The great popularity of *El Lazarillo de Tormes* can be seen in Chapter XXII of *Don Quixote* (1605) when the character Ginés de Pasamonte claims that his own autobiography would surpass its success. The picaresque novel, Cervantes noted, had become a vehicle for self-expression available to anyone who could write. Ginés de Pasamonte realizes that his own story could become a valuable asset in the newly created market of printed books. Within fifty years (1554–1605), self-expression had become a form of personal empowerment and writing had become a way of creating value. Thus, as another character informs us, Ginés de Pasamonte pawns his manuscript for 200 *reales*. *El Chavo del 8* became immensely popular in the 1970s mainly due to three similar circumstances. First of all, Latin America, and especially Mexico—with its population of almost 50 million at that time,[1] had undergone a major transformation from rural to urban societies experiencing demographic explosions that created megacities such as Mexico City, Bogota, and Buenos Aires. Secondly, the consolidation of television networks throughout the region led to the formation of mega-companies like *Televisa* (1973) that used the new VCR technology to create and distribute videotaped copies of TV shows for repeated broadcasting in Spanish-speaking countries. Thirdly, the strengthening of a regional market for cultural products in which Latin American identity was broadly defined, by producers like *Televisa* itself, using the Spanish language and their shared colonial experience as elements of differentiation from the hegemonic culture of the United States.[2] Thus, for Latin American audiences, the TV shows created by Roberto Gómez Bolaños, better known as Chespirito, became both counterparts and competitors of popular North American series like *The Carol Burnett Show* (1967–1978)[3] and *I Love Lucy* (1951–1957) which were dubbed into Spanish and routinely transmitted throughout the region. In Latin America, the national hegemonies had to adapt to the new transnational public space established by the audiovisual mass media while maintaining, or attempting to maintain, ideological preeminence within their own nations.[4] With the exception of Cuba, that had undergone a revolution in 1959, and Chile with its socialist government between 1970 and 1973, the countries of Latin America opted for a sort of mediated relationship with the U.S. model that was based mainly on the so-called "invisible hand of the market."[5] As an alternative, and as I have argued in my book on *El Chapulín Colorado*, Latin American hegemonies have offered a hybrid between feudalism and capitalism, a sort of feudal-bourgeois system, articulating residual ideologies and relations inherited from the Spanish Colonial Era with the ideas produced in the new transnational market for cultural products. Such residual ideas and their effectiveness can be identified in audiovisual products that now form part of what Raymond Williams (1998) calls the "documentary" culture as it records the human experience at a given moment in history (p. 48). Unveiling such residual ideologies is the first step in empowering the readers, spectators and consumers of such products, permitting them to develop a historical and dialectic approach to contemporary popular culture.

Content wise, the picaresque novels and *El Chavo del 8* discuss a variety of common themes such as childhood, schooling, inner-city poverty, child labor, and social structures. An analysis of the parallels and equivalences among these topics will reveal the extent to which *El Chavo del 8* inscribes itself within the picaresque literary tradition, adapting it to new forms of production and representation used in audiovisual media. In the following pages, I will discuss such parallels and correspondences between the *El Chavo del 8* TV series and four picaresque novels ranging from the mid-sixteenth century to the mid-twentieth century: *El Lazarillo de Tormes* (1554), *El Buscón* (1626), *El Periquillo Sarniento* (1819), and *La vida inútil de Pito Pérez* (1938). The first two were written in Spain during the so-called Golden Age of the Spanish Empire and have long been a part of the literary canon in Spain and Latin America. The last two are set in Mexico during the colonial and postcolonial periods and are forerunners of *El Chavo del 8* in representing the struggle of working-class characters in urban settings.

The name and its origin

The Spanish word *chavo* is used in Mexico to refer to a young boy who has not yet reached puberty. It is not a given or legal name but simply a nickname. In Spain, which is often called *la madre patria* or "motherland" in Latin America due to its imperial reign over the region for more than three centuries, the term *chavo* is derived from *ochavo*, an old name for coins of very low value, somewhat like the American penny nowadays. As such, the nickname may originally have had clear social class implications most likely inherited from the time when

Mexico, under Spanish rule, was called "New Spain." Nonetheless, El Chavo's legal name is never mentioned in any of the 290 episodes of the 1970s TV series and the sketches that followed it. The truth is that Roberto Gómez Bolaños[6] named the TV program *El Chavo del 8* because the show was originally aired on Channel 8. Later, when *Telesistema Mexicano* and *Televisión Independiente de Mexico* merged to form *Televisa* in 1973, the show was moved to Channel 2,[7] but the name remained the same. On various occasions throughout the series it becomes clear that El Chavo does not live in a barrel but in apartment number 8, while the barrel is, so he says, his secret hiding place. On several occasions when other characters inquire about his real name, El Chavo attempts to respond but is always interrupted and the audience never gets to hear his answer, so the TV show provides very few clues to his family background. However, Roberto Gómez Bolaños published a book titled *El diario de El Chavo del Ocho* in 1995 in which he clarifies certain biographical details of El Chavo which never appeared in the TV series. The book includes a preface in which Bolaños, following Cervantes' model in *Don Quixote*, says that he met a shoeshine boy called El Chavo del Ocho[8] at a park. After being paid for his work, the boy ran off, leaving a notebook behind on the bench, which, according to the narrator, contained the text of the book. In the preface, the historical author, RGB, also presents himself as the editor and publisher of El Chavo's writing and hides behind one of his own characters to tell El Chavo's open-ended story in first person. Thus *El diario de El Chavo del Ocho* has a textual typology proper to the picaresque novel and, except for the main character never coming of age, it displays all the characteristics of the subgenre.

With respect to his family, El Chavo tells us in the book that he did not know his father because he "*nomás se acostó y se fue*"/"He only went to bed [with his mother] and left" (p. 11). He did know his mother but "*nomás tantito*"/"only a little." For unknown reasons, she didn't come to pick him up from daycare one day and suddenly he became either an orphan or an abandoned child. After living at an orphanage for a while, the boy ran away, winding up in a *vecindad* (a privately owned block in a working-class neighborhood divided into dwellings for low-income tenants) where he finds shelter with "*una señora muy viejita*"/"a very old woman" in apartment number 8. When the old lady dies, he becomes a homeless boy who now relies on his friends who take turns in offering him shelter (p. 27). Thus, according to El Chavo himself, the nickname is derived from the place where he first stayed when he arrived in the *vecindad*.

El Chavo's personal story is nothing new. In fact, I would argue that it belongs to a longstanding literary tradition that originated in Spain more than 400 years ago with the publication of *El Lazarillo de Tormes* (1554), in which Lázaro, the town crier of the city of Toledo in Spain, recapitulates his life, telling us the story of his upbringing in first person. Regarding his name, he says:

> *a mí llaman Lázaro de Tormes, hijo de Tomé González y de Antona Pérez, naturales de Tejares, aldea de Salamanca. Mi nacimiento fue dentro del río Tormes, por la cual causa tomé el sobrenombre* (p. 12).

> [. . .] my name is Lázaro de Tormes, son of Tomé González and Antona Pérez, natives of Tejares, a village near Salamanca. I was born in the River Tormes, which is why I took that nickname . . . (My own translation).[9]

Nowadays in Spain, where paternal and maternal last names are used together, his legal name would be Lázaro González Pérez, but we know him by the combination of his given name and the toponym of the river in which he was born. When he is only eight years old, the same age as El Chavo, Lázaro loses his father and is left in the care of his mother who gives him away to a blind man to be his "boy," guide and servant.

There are several parallels between Lázaro de Tormes and El Chavo del Ocho: their names do not derive from their biological parents but from place names; their various trades and occupations; and the early age at which they are forced to fend for themselves. However, while Lázaro grows up in the novel, learns new trades and eventually gets married, *El Chavo del 8* has a floating timeline that keeps the character frozen in time as a perpetual eight-year-old. In this sense, *El Chavo del 8* follows a narrative procedure similar to that of the animated TV series *The Simpsons* in which the characters do not age throughout the course of 27 seasons. Maintaining this floating timeline allows RGB to update and rewrite his characters without having to represent their aging process and the social consequences of such impoverished living conditions and upbringing. His characters have no future, reality never varies for them, and thus they never face the sociopolitical consequences of the injustices represented in both the book and the TV series.[10]

In the case of *El Periquillo Sarniento* (1819), the Mexican picaresque novel by José Joaquín Fernández de Lizardi, the main character's name is altered several times: first, from Pedro to Pedrillo; later, because he attended school wearing "*chupita verde y calzón Amarillo*"/"a green jacket and yellow pants"; and finally, because he suffered from scabies (19–20), his classmates nickname him Periquillo Sarniento, which has been translated into English as "Mangy Parrot."[11] In the novel *La vida inútil de Pito Pérez* (1938), written in Mexico by José Rubén Romero, the protagonist gets his name from a type of flute known as *pito de carrizo* that he made with a pocket knife. Since he was always seen or heard playing this flute, people begin calling him Pito Pérez (24). As an adult, Pito Pérez is given another nickname that he dislikes because it reminds him of society's contempt and rejection of him. They call him Hilo Lacre, a nickname derived from his employment as *barrilero*, a sort of roaming street vendor. His flute, he says, was most likely lost in a jail cell during one of his many arrests. In his testament, Pito Pérez complains openly and denounces society for mistreating him:

> *[Humanidad] ¡Hasta de mi nombre me despojaste para convertirlo en un apodo estrafalario y mezquino: Hilo Lacre!* (p. 92).
>
> [Humanity] You even deprived me of my name to turn it into a bizarre and petty nickname: Hilo Lacre!

To Pito Pérez, the nickname becomes a stigmatizing mark that society imposes on him. A great part of his identity is simply imposed on him by "humanity," which becomes a metaphor for the social structure that subjects him to its disdain. The main characters in the picaresque novel always reject such social impositions and seek ways to elude them.

Changes of names and their frequent alteration are common practices in picaresque novels. In the case of Francisco de Quevedo's *El Buscón* (1626), the main character—Pablos—changes his name as part of a strategy to hide his origins and gain social acceptance. In his early adulthood, he presents himself as Don Felipe Tristán (p. 264); by using this name he hopes to receive preferential treatment in new social circles. Pablos wants to project his personal volition and desires in the form of a social identity around him. In his struggle for social mobility he does not want people to associate him with his parents—considered criminals at the time—nor with his uncle, an executioner. As a child in school, some of his classmates would call him Don Razor or Don Swindler to remind him that his father was a barber and a thief, and his mother a whore and a witch. Surnames refer to a person's family origin and in stratified societies such as those of Spain and Latin America, they determine the way someone will be treated by others. In feudal-like societies, a politically connected surname or one related to traditionally powerful families is often more important than academic degrees and work experience, especially for positions in the public sector, where nepotism is a common practice. In the case of El Chavo, his name makes him yet another homeless child among many; he is a *peladito*—a penniless urban waif belonging to a generation descending from the *rascuache* identity of Cantiflas.[12] In other words, he comes from the great masses of lumpen proletariat that have been moving to the semi-industrial cities of Latin America in search of work since the first decades of the twentieth century. Like the great majority of people in Latin America, El Chavo has no idea who his ancestors were and has been dispossessed of any material belongings or historical memory of his family, practically since birth. In contrast to Ñoño—the son of Mr. Barriga, El Chavo will never inherit any income-producing properties to guarantee his livelihood, nor will he ever be able to count on his father's friends and contacts to find a job.

Teachers, school, child labor, and physical abuse

Another theme that *El Chavo del 8* shares with the picaresque novel is the representation of childhood and schooling. Sometimes, as in the case of *Lazarillo de Tormes*, a child's instruction and labor are mixed into one process. As noted previously, Lázaro is given in servitude at the age of eight to The Blind Man who would supposedly show him "how to make a living" in exchange for his services (28).[13] According to Lázaro, The Blind Man was astute and cunning and "had endless ways of getting money out of people" (p. 28).[14] Generally speaking, The Blind Man was a mixture of healer and prayer for hire who:

> *Decía oraciones para muchos y diversos efectos: para mujeres que no parían; para las que estaban de parto; para las que eran malcasadas, que sus maridos las quisiesen bien. Echaba pronósticos a las preñadas si traían hijo o hija* (p. 26).
>
> [. . .] knew prayers for lots of different things: prayers for women who couldn't have children, prayers for women who were pregnant, and prayers for women who were unhappily married, to make their husbands love them; he would also tell pregnant women if they were going to have a boy or a girl. (p. 28)

He also treated toothaches, fainting fits, morning sickness, and other ailments with boiled herbs and roots. The skills that Lázaro learned during his time with The Blind Man allow him to find employment as an assistant to a stingy priest, as a servant to a penniless squire, as assistant to a seller of indulgences, among other occupations, and finally as a town crier in Toledo. Lázaro never spends time with children his own age because he has to earn a living from the beginning in a cruel and miserable society. Like Lázaro, Pito Pérez (1938) is "educated" on the job: first, he is an altar boy in church; later, he works at his uncle's grocery store; he also works with an apothecary learning how to prepare "capricious prescriptions" made with "simple syrup and harmless cornstarch pills" (pp. 29–30).

On the other hand, Pablos (*El Buscón*) and Periquillo Sarniento attend school and narrate their experiences there. Pito Pérez certainly receives some formal instruction, but it is not a main part of the narrative. Periquillo Sarniento, who narrates his story from the perspective of a grown man wishing to leave his testimony so his own children and any other young readers can learn from his experiences, assumes a moral tone and criticizes individuals, including himself, as well as institutions. So he tells us that he first went to school when he was about six years old, and he immediately denounces his teacher's professional shortcomings. Periquillo recalls hearing the man say "*sólo la maldita pobreza me puede haber metido a escuelero*" (p. 14)/"It was just this damned poverty that turned me into a schoolteacher." Teaching was neither a choice nor a vocation for him, but rather the only alternative given his socioeconomic situation. Periquillo also complains about his dreadful lack of teaching skills:

> *[M]i maestro carecía de toda la habilidad que se requiere para desempeñar este título. Sabía leer y escribir, cuando más, para entender y darse a entender; pero no para enseñar* (15).
>
> [M]y teacher lacked all of the skills required to perform as such. He knew how to read and write just enough to understand others and to make himself understood, but not enough to teach.

Periquillo is also adamant in recalling that his first teacher knew some calligraphy but none of the rules of punctuation or spelling. The school, he says, was not only poor, but badly directed. The teacher had no classroom-management skills and was seen as soft by the students, including Periquillo, who really ruled the place. This first teacher is eventually forced to close the school and Periquillo is transferred to another institution where he is under the tutelage of a teacher he describes as:

> *[H]ombre de bien a toda prueba, arrogante lector, famoso pendolista, aritmético diestro y muy regular estudiante; pero todas estas prendas las deslucía su genio tétrico y duro* (p. 24).
>
> A good man, an arrogant reader, with great penmanship, skillful at arithmetic, a disciplined student; but all those talents were overshadowed by his gloomy and ill temper.

Physical punishment was common practice in school, and Periquillo denounces the fact that his teacher corrects his mistakes by whipping him (p. 25). Thanks to his mother's intervention, Periquillo is transferred to another school after just a couple of months. For the next two years he is educated by a man whose face "*manifestaba la dulzura de su corazón*"/"showed the sweetness in his heart," a teacher who would only use physical punishment in extreme cases (p. 29). Once he has learned how to read and write, Periquillo becomes a student of Latin grammar under the tutelage of a reputable teacher: Don Manuel Enríquez. In his account of his time as a student of Latin grammar, he confesses that he was "*el muchacho más maldito entre los más relajados estudiantes*"/"the trickiest and most laid-back among the students," and he also considered himself to be "*non plus ultra de los bufones y chocarreros*"/"non plus ultra among the clowns and truants" (p. 48). Nevertheless, he completes his studies and graduates with a Bachelor of Arts from *Colegio de San Ildefonso* (p. 54). Although the book does not go into specifics, the narrator informs that the teachers in the school always had a "cane" and "donkey ears" to impart physical punishment and humiliation on the students. In summary, Periquillo seems to advocate a middle ground between his first and his second school experience that is dialectically resolved in the figure of his third teacher, who combines vocation, presented as sweetness of heart, and ruling by fear in the form of physical punishment to address "extreme cases."

Among the four picaresque novels discussed here, *El Buscón* (1626) has the most similarities with *El Chavo del 8* on the topic of schooling. In Quevedo's picaresque novel, Pablos enters school with the intention of learning how "to be an honest man" (p. 87)[15] because he wanted to "go forward with his good ideas."[16] In other words, Pablos wants to leave behind his precarious origin and attain social mobility. In his first school, he finds friendship and rejection, esteem and contempt. In reading his story, we soon learn that due to his obsequious attitude towards his teacher and the teacher's wife, on many occasions he is put in charge of the cane to whip his classmates when instructed to do so. This position as the one who dispenses punishment sometimes led to rejection and sometimes to many favors from "*los hijos de caballeros y personas principales*"/"the sons of gentlemen and principal people" (p. 95). Although El Chavo is never put in charge of punishing other students in school like Pablo was, he is always expected to be helpful and to assist the adults around him. In his first school, Pablo meets Don Diego Coronel, his playmate and future friend. Thanks to this friendship, Don Diego's family invites Pablos to accompany him as his servant and attend boarding school in Segovia under the tutelage of Dr. Goat. Childhood friendship is thus transformed into a type of servitude that allows Pablos to continue his education. El Chavo is invited to Ñoño's house, to Quico's parties, to Acapulco with Mr. Barriga, and to breakfast at La Chilindrina's in the same way: always as a friend, companion, and someone who has to be of service to others. Like Pablos, El Chavo attends school in the company of those who provide him with shelter and food. For Pablos and Don Diego, the boarding school experience is marked by starvation and scarcity. Pablos is usually so hungry that, he would "eat" half of the phrases when asked to repeat a lesson, that is to say, he makes no sense (108).[17] Their experience at the boarding school ends when a classmate dies of starvation and Don Alonso Coronel, Don Diego's father, takes them away from the place. In *El Chavo del 8,* no one starves to death, but the main character is always hungry and instead of bringing an apple to his teacher on the day of the final exams, he brings the leftovers of several already eaten apples and even takes another apple from the teacher's desk. Pablos later goes with Don Diego, once again as his servant, to Alcalá de Henares where he endures the mockery and attacks of students and their servants who spit at and beat him as part of a welcoming party. Don Diego, on the other hand, uses his money to rescue himself and is given the treatment and privileges of an old mate by the other students who receive the money and yell:

> *¡Viva el compañero, y sea admitido en nuestra amistad. Goce de las preeminencias de antiguo. Pueda tener sarna, andar manchado y padecer la hambre que todos* (p. 123).
>
> Long live our comrade, admit him to our circle! Let him enjoy the privileges of an old student! Let him have the pox, go unwashed, and get hungry like the rest of us! (p. 106)

Picaresque novels usually represent students both in Spain and in Latin America as dirty, unhealthy, and always hungry. This representation is obviously a stereotype that most likely developed in the sixteenth and seventeenth centuries in urban centers built around universities such as Alcalá de Henares and Salamanca in Spain. In *El Chavo del 8,* El Chavo is the only one who seems to be lacking in hygiene. In an episode now known as "*Al agua Chavo,*" his teacher and his neighbors take on the mission of giving him a bucket bath.[18] Like Don Diego in *El Buscón,* in *El Chavo del 8* other characters like Ñoño, Quico, and La Popis also attempt to buy the favors and friendship of other children and the teacher by offering them presents such as melons, apples, candy, or letting them play with their toys. Furthermore, Quico and La Popis attempt to manipulate the teacher's romantic interest in Doña Florinda (known in English as Mrs. Worthmore) to obtain special treatment in school.

In his perpetual childhood, El Chavo del Ocho has similar experiences to those of the *picaros* discussed here. Like Lázaro de Tormes and Pito Pérez, El Chavo is constantly taking on odd jobs that allow him to survive and learn a trade. Sometimes we see him running errands in the *vecindad*, shining shoes, selling lemonade with his friend Chilindrina, washing Mr. Barriga's car, helping Mr. Ramón in his work as a photographer, and serving as the only employee in Doña Florinda's restaurant. Towards the end of the TV series in 1979, there is an episode called "*Manifestación por El Chavo*"/"A Demonstration for Chavo" in which Mrs. Nieves organizes and leads a protest march fighting for El Chavo's rights as a worker. In this episode, the series acknowledges the reality of child labor in Latin America and advocates the basic rights of young workers. The episode ends with a scene in which Doña Florinda tells El Chavo that from then on he will receive a salary and have the right to eat as much as he wants outside of his working hours. As mentioned before, the floating timeline keeps the events represented in an everlasting present that permits RGB to rewrite, or at least to attempt to rewrite, some of the episodes. Nonetheless, after almost a decade on the air, this is the only episode in the series to represent the transition between unpaid child labor and work for wages. Child labor was there to stay and the TV series would not advocate an end to it in this or any other episode, but simply an attenuation of its harsh conditions. This is the closest the series would come to portraying El Chavo as achieving new social status as a paid worker. If we look back to *El Lazarillo*, we can see that El Chavo now joins Lázaro who slowly moves away from the constant threat of starvation to stable and secure employment.

In *El Chavo del 8* and the four picaresque novels discussed here, child labor is mixed with acts of physical punishment represented as normal practices in their respective times and societies. The Blind Man beats Lázaro repeatedly. One day, upon discovering the boy drinking from a jug without his permission, he hits him with it so hard that Lázaro recalls the events saying:

> *Fue tal el golpecillo, que me desatinó y sacó de sentido, y el jarrazo tan grande, que los pedazos dél se me metieron por la cara, rompiéndomela por muchas partes, y me quebró los dientes, sin los cuales hasta hoy día me quedé* (p. 33).

> The blind man's little tap was so hard that I was knocked right out and had bits of broken jug stuck in my face and was cut all over. My teeth were broken and that's why I haven't got any in my head today (p. 31).

Pito Pérez receives similar treatment from Father Coscorrón[19] who discovers that he and Dimas, another altar boy at the church, have been stealing from the poor box. The *pícaro* recalls the punishment:

> *El cura agarró con sus dedos de alambre una de mis orejas, que poco faltó para que se desprendiera de su sitio* (p. 22).

> The priest grabbed my earlobe with his wiry fingers so hard that he almost ripped it off.

It could be argued, following Gerald Mast's (1979) definition of the so-called "comic climate," that since *El Chavo del 8* is a comedy, it is set in a climate in which the violence represented is considered trivial and unimportant, and is therefore seen as simply harmless comical action. El Chavo himself accidentally hits Mr. Barriga every time the man comes to collect the rent from his tenants. The collective fights portrayed in the series, which sometimes involve cream cakes, can also be interpreted in a Bakhtinian sense as carnivalesque. Nevertheless, the adults in the series, especially Mr. Ramón, usually express their frustration with El Chavo by hitting him, mostly to punish him for his supposed insolence. The most common example of this occurs every time Doña Florinda hits Mr. Ramón and tells him to go deal with his grandmother. On such occasions El Chavo immediately asks about his grandmother's likes and dislikes, at which point the frustrated angry man invariably hits the child and makes him cry. Mr. Ramón also frequently pinches Quico on the arm when he is frustrated either with the boy or his mother. Therefore, although we do have to accept that some of the violence represented in the series corresponds to the comic climate, I would argue that some forms of physical punishment, specifically hitting El Chavo and pinching Quico, are elements of verisimilitude that connect the series with the literary tradition of the picaresque novel. They also reveal how Latin American societies in the 1970s and 1980s still condoned such violence as normal or acceptable behavior. The reports of the Global Initiative to End All Corporal Punishment of Children[20] show how corporal punishment of children still prevails in Mexico and Latin America at large. In fact, as of September 2015 in Mexico, the law did "not explicitly prohibit all corporal punishment in childrearing" (p. 2). In other words, these actions could be read as a verisimilar representation of the living conditions of many children in the 1970s and 1980s and not merely as moments of slapstick comedy.

As shown above, when discussing schooling, the main characters in the picaresque novels *El Buscón* and *El Periquillo Sarniento* are subjected to institutionalized physical punishment inflicted by their teachers and/or classmates. On the contrary, in *El Chavo del 8,* school is physically, if not emotionally, the safest place for children to be. School is a place that is mostly void of corporal abuse. However, on one occasion, in the episode entitled "*De regreso a clases*"/"First Day of School," Mr. Ramón joins El Chavo's class and pinches Quico in the arm in front of his classmates and the teacher does nothing to condemn his actions. There is another episode that includes the infamous "donkey ears" that are not assigned to a child in this case, but to Mr. Ramón who once again comes to visit the school and decides to stay for class. Profesor Jirafales is presented in the TV series as an improved and somewhat progressive version of Periquillo Sarniento's third teacher who, according to Periquillo, was:

> *[S]emijoven como de treinta y dos a treinta y tres años, de un cuerpo delgado y de regular estatura; vestía decente, al uso del día y con mucha limpieza; su cara manifestaba la dulzura de su corazón en una palabra, este hombre amable parece que había nacido para dirigir la juventud en sus primeros años.* (p. 28)
>
> Youngish, about thirty-two or thirty-three years old, skinny and of average height; decently clothed, in style and very clean; his face showed the sweetness of his heart. In a word, this man seemed to have been born to guide young people though their first years.

Periquillo's third teacher makes it known that he has instruments to impart physical punishment in the school, but he keeps them locked away because they are "*instrumentos horrorosos que anuncian el dolor y la infamia*" (p. 29)/"horrendous instruments that announce pain and infamy." Such instruments were not created, he clarifies, for children attending this school because they do not come from ordinary (i.e., low-class) families. As an educator, Profesor Jirafales tries to treat his students properly, lecturing and taking questions from most of the students.

In the series, Profesor Jirafales, along with Jaimito the Mailman, represents the image of Latin American public servants and the government in general. These characters reveal the internal institutional contradictions of Latin American nations in the 1970s and 1980s. Profesor Jirafales is the educator who stands in front of a classroom filled with students from a variety of social classes. His class includes: the son of an upper middle-class bourgeois, the child of an impoverished widow that he dates, the daughter of a chronically unemployed (lumpen proletariat) bachelor, and El Chavo, a nameless orphan. Despite his constant failures at teaching, Profesor Jirafales is persistent in attempting to educate such a diverse group of students. Like Sisyphus, he faces a seemingly impossible task, but he goes back to the classroom every day and tries to teach again and again. Social reality filters into his classroom, as El Chavo arrives hungry everyday dreaming of a ham sandwich. Jaimito the Mailman, and his desire to avoid all effort, represents the slothfulness of bureaucracy in public institutions that resist efficiency at all costs. Jaimito is so slow that private citizens often end up delivering the mail for him. These two characters seem to represent the need for basic public services, but they also serve to denounce the incapacity of some public institutions to fulfill their missions. In other words, *El Chavo del 8* subtly argues in favor of private or semi-private institutions to replace the inefficient parts of Latin American governments. In the same fashion, the picaresque novels discussed here present open and subtle criticisms of institutions like the church, in *El Lazarillo*, and schools and universities in the other three. Like El Chavo del Ocho, Lázaro and one of his masters, the Squire, rely on private charity to survive in Toledo. Lázaro begs for food while his master pretends to be wealthy and courts women in his search for a profitable marriage.

Social services are notoriously absent in *El Chavo del 8*. Instead of a public welfare system to provide for his care, the series shows the neighbors (*vecindad*) assuming this responsibility. Organic, collective action replaces the state by default. Sometimes El Chavo has to earn money like Lázaro or Pito Pérez, by running errands or serving the neighbors, while on other occasions he depends on acts of charity. Mr. Barriga takes him to Acapulco a couple of times for vacation. Like the four picaresque novels discussed here, *El Chavo del 8* documents the inner functioning of societies in urban settings. Many social functions that were performed by the church in the time of *El Lazarillo*, *El Buscón*, and *Periquillo Sarniento*, later became the responsibility of the state in the postcolonial national projects of the twentieth century. Pito Pérez and El Chavo both represent children educated within the national project, and their stories show the limits and inefficiencies of public institutions. While there are plenty of jails to imprison Pito Pérez, there are no social services to provide food and shelter for El Chavo once he runs away from the orphanage. Furthermore, no one inquires into the matter of his being abandoned and everybody takes his situation as something quite normal in its context.

In conclusion

In 1554 *Lazarillo de Tormes* initiated the long tradition of the so-called picaresque novel, a type of *bildungsroman*, or coming-of-age narrative that has spread throughout the world from that time to the present day. Its great popularity, followed later by that of *El Buscón* (1626), *El Periquillo Sarniento* (1819), and *La vida inútil de Pito Pérez* (1938), among many others, can only be compared with the immensely successful audiovisual production of *El Chavo del 8* in the second half of the twentieth century. Furthermore, the character of El Chavo del Ocho is built upon the tradition of the picaresque novel both in content and form in the written version. However, *El Chavo del 8* was intended to be a realistic audiovisual comedy dealing with issues of childhood and inner-city poverty on a never-ending floating timeline. In other words, the characters are not allowed to grow old and therefore we can only trace the parallels and equivalences between El Chavo and the childhood and education of the *pícaros*. The picaresque novel has always included severe criticisms of institutions that provided social services, such as the Catholic Church and schools. Moreover, as in the case of *El*

Buscón, they have focused on social structures and emergent groups that threaten to disrupt the fragile stability of the social pyramid. *El Chavo del 8* includes a veiled criticism of public institutions, but it does not openly condemn them. With the exception of schools and the inefficient mail service, public services are not depicted at all in the show and, except for the orphanage, they are not mentioned in the book *El diario del Chavo del Ocho*. The existence and responsibility of the state is limited; private citizens are left to fend for themselves and develop forms of communal collaboration or charity. On the other hand, like *El Buscón* and *Periquillo Sarniento*, the series represents the school experience of the main characters while parodying the almost impossible task of teaching, as it is bravely undertaken by Profesor Jirafales. *El Chavo del 8* also depicts child labor and physical abuse in an allegedly comic atmosphere, which I nonetheless argue is as harsh and real for El Chavo as it was for Lázaro, Pablos, Periquillo, and Pito in their day.[21]

Notes

1 See *Estadísticas históricas de México Tomo II* (1999). Available at: www.inegi.org.mx/prod_serv/contenidos/espanol/bvinegi/productos/integracion/pais/historicas/EHM%201.pdf

2 For an extended discussion of this process see the chapter "Nación" in my book *¡No Contaban con mi astucia! México: parodia, nación y sujeto en la serie de* El Chapulín Colorado (2014) (pp. 141–164).

3 This show ran for 11 seasons and had 279 episodes.

4 In 1977, Armand Mattelart published *Multinacionales y sistemas de comunicación* revealing the conditions of the telecommunications market at the time. There, he shows that in 1972, an average of 60–70% of all TV aired in Latin America was imported from the United States (p. 248). He also shows how many Latin American political campaigns were coordinated and strategized from New York (pp. 270–279).

5 For an analysis of the how globalization has affected Latin American television, see John Sinclair's (1999) seminal book *Latin American Television: A Global View*.

6 For a history of radio and TV in Mexico and a discussion of RGB's career as a writer and actor, see "Precisiones históricas" in my book *¡No Contaban con mi astucia! México: parodia, nación y sujeto en la serie de* El Chapulín Colorado (pp. 44–63).

7 Channel 2 had wider broadcasting power than Channel 8.

8 When mentioning the TV series I will use "El Chavo del 8" and when discussing the character I shall use "El Chavo del Ocho" or simply "El Chavo."

9 In his translation, Michael Alpert translated "*sobrenombre*" as a surname, but a more accurate translation would be "nickname."

10 This concept of an "invariable reality" is taken from *Para leer al Pato Donald* (*How to Read Donald Duck)* by Ariel Dorfman and Armand Mattelart (p. 25).

11 It might be pure coincidence but both green and yellow are also present in El Chavo's costume. His hat and his short pants (especially in the animated series) have a clear resemblance to Periquillo's attire.

12 For a discussion on *rascuachismo*, see "The Riddle of Cantinflas" by Ilan Stavans (1995).

13 In the original "*adiestra en la carrera de vivir*" (p. 24).

14 In the original "*mil formas y maneras para sacar el dinero*" (p. 26).

15 In the original "*aprender virtud resueltamente*" (p. 93).

16 In the original "*buenos pensamientos adelante*" (p. 93).

17 In Spanish "*comerse las frases*."

18 The episode, remade several times, begins with a moral tone which evolves into a carnivalesque water battle that ends with everybody soaking wet.

19 In the original "*padre Coscorrón*"; that literally means Father "Smack on the Head."

20 Available at: www.endcorporalpunishment.org.

21 What would the adulthood of El Chavo be like? If we were to follow the tradition of the picaresque novel in the work of Roberto Gómez Bolaños, I would argue that El Chavo's adult life is more or less represented by the character of El Chómpiras from the sketch called *Los Caquitos* (The Little Thieves) that ran from the 1970s to the mid-1990s.

References

Aguasaco, C. (2014). *¡No Contaban con mi astucia! México: parodia, nación y sujeto en la serie de* El Chapulín Colorado. Monterrey: Universidad Autónoma de Nuevo León.

Anónimo. (1554). *Lazarillo de Tormes* (2000). Madrid: Cátedra.

Anonymous and Quevedo, F. De. (1554/1626). *Two Spanish picaresque novels: Lazarillo de Tormes and The Swindler.* Trans. Michael Alpert. New York: Penguin Books.

Bajtin, Mijail. (1998). *La cultura popular en la edad media y en el renacimiento.* Trans. Julio Forcat and César Conroy. Madrid: Alianza Editorial, 2005.

Cervantes Saavedra, Miguel de. (1605). *Primera Parte del Ingenioso Hidalgo don Quixote de la Mancha.* Madrid: Cervantes Virtual.

Dorfman, A. and Mattelart, A. (1972). *Para leer al Pato Donald.* Buenos Aires: Siglo XXI.

Estadísticas históricas de México Tomo II (1999). Mexico DF: Instituto Nacional de Estadística, Geografía e Informática.

Fernández de Lizardi. J.J. (1819). *El Periquillo Sarniento* (1842). Mexico: Librería de Galván.

Global Initiative to End All Corporal Punishment of Children (2015). "Corporal punishment of children in Mexico." Available at: www.endcorporalpunishment.org/assets/pdfs/states-reports/Mexico.pdf.

Gómez Bolaños, R. (1971–1980). *El Chavo del 8* [Television series]. Mexico DF: Televisión Independiente de México & Televisa.

Gómez Bolaños, R. (1995). *El diario de El Chavo del Ocho* (2005). Mexico DF: Punto de Lectura.

Mast, Gerald. (1979). *The comic mind: Comedy and the movies.* Chicago and London: The University of Chicago Press.

Mattelard, Armand. (1977). *Multinacionales y sistemas de comunicación.* Mexico DF: Siglo XXI.

Quevedo, F. De. (1626). *La vida del Buscón llamado Don Pablos* (2000). Madrid: Cátedra.

Rodríguez, J C. (1990). *Teoría e historia de la producción ideológica: las primeras literaturas burguesas (siglo XVI).* Akal universitaria. 2nd edn Madrid, España: Akal.

——. (2002). *Theory and history of ideological production: The first bourgeois literatures (the 16th century).* Trans. Malcolm K. Read. Monash Romance Studies. 1st American ed. Newark [Del.]: University of Delaware Press.

Romero, J.R. (1938). *La vida inútil de Pito Pérez* (2007). Caracas: Fundación Editorial el perro y la rana.

Sinclair, J. (1999). *Latin American television: A global view.* Oxford and New York: Oxford University Press.

Starr, P. (2004). *The creation of the media: Political origins of modern communications.* New York: Basic Books.

Stavans, I. (1995). 'The Riddle of Cantinflas.' *Transition,* 67, 22–46.

Williams, Raymond (1998). 'The Analysis of Culture'. In *Cultural theory and popular culture.* Ed. John Storey. Harlow: Prentice Hall, 48–56.

——. *Marxism and literature* (1977). London: Oxford University Press.

5

Border-crossing Chespirito

El Chavo del 8 Meets Pepito in Exile

Limarys Caraballo

The unforgettable *Chavo del 8* and *Chapulín Colorado*, created and delivered by writer, actor, and director Roberto Gómez Bolaños, are woven into the fabric of my childhood memories. As a Cuban and Puerto Rican American born just outside of San Juan in the 1970s, for me the personages of *El Chavo* and *El Chapulín* were on par with Tom and Jerry, Mighty Mouse, and Wonder Woman. While I enjoyed the shows, my parents seemed to look forward to them as much as I did, and the characters' antics became part of our family's humorous repertoire. If I sheepishly told my mother that I didn't mean to leave crumbs on the furniture after my snack, she would joke that I did it "*sin querer, queriendo*" [without meaning to] like El Chavo, the bumbling orphaned boy played by the adult Gómez Bolaños, who exasperated and yet tugged at the heartstrings of the other residents of the *vecindad* [tenement housing]. If my father managed to fix a favorite toy, he might smile triumphantly and exclaim, *¡No contabas con mi astucia!* [You underestimated my intelligence!] to invoke Gómez Bolaños' kind-hearted—but fairly inept—underdog superhero, Chapulín Colorado.

Reflecting on Chespirito's prominence and the international and intergenerational nostalgia evoked by his passing in 2014, I wondered about the role of characters like El Chavo in comparison to other cultural icons. For me, given my Cuban heritage and Puerto Rican background, El Chavo's antics were just as familiar as those of Pepito, a mischievous personage in many Latin American jokes and tales.[1] Like Chavo, Pepito is recognized and cherished in many Spanish-speaking countries: a boy that never grows up, whose escapades inspire laughter and incite social commentary. In contrast, however, to the appeal of Chespirito's sketches as wholesome family programming, Pepito is the protagonist of countless off-color jokes and tales about politics and double entendre. A precocious and mischievous elementary school boy, Pepito constantly foils his parents, teachers, principals, and even political leaders' efforts to maintain order and propriety. Like similar fictional figures known throughout Latin America and Spain as Benito or Jaimito, the version of Pepito that features in many Cuban jokes and stories (whether on or off the island), gets himself into all kinds of scrapes and is associated with jokes that reflect a nation's (and a nation-in-exile's) widely held values. It is the Cuban version of Pepito, as frequently featured in jokes about socioeconomic, political, and educational issues on the island, that I juxtapose with the character of Chavo, who is much less overtly connected to political and social agendas in his own country of origin. Juxtaposing these two iconic characters, who are both similar and divergent in many ways, supports a framing of the humorous border crossings of Chespirito and Pepito as sites of possibility for rethinking static notions of identities, cultures, and citizenship (Caraballo, 2011, 2012; Guerra, 2012, 2015) with/in immigrant and minoritized communities.

In this chapter, I examine a set of narratives about El Chavo and Pepito gathered from members of various generations among my family and friends from the Caribbean and Latin America, as well as my own recollections. Based on these narratives, I frame Chavo and Pepito as situated in particular cultural worlds, described by Holland et al. (1998) as *figured worlds*. Reading across our reflections about Chavo and Pepito as cultural icons, my inquiry underscores the values and cultural perspectives implicit in these jokes and sketches that so many have enjoyed for generations. Framing humor as socially constructed and culturally situated (Alharthi, 2014), I examine how the resonances of Chavo and Pepito reflect the identities and experiences of their audiences, as these popular icons cross national and cultural borders and assume greater sociopolitical significance in the context of current discourses and debates about identity, culture, and citizenship.

Humor and identities in transcultural contexts

According to Holland and colleagues (1998), people look at the world and themselves from the positions in which they are persistently cast, as well as drawing from an existing lexicon of culturally produced types (such as, but not limited to, nationality, gender, race, ethnicity, etc.). In other words, we construct our own identities and understandings of others' identities by building upon the "narratives" that are available in any given cultural or figured world. A figured world is a "realm of interpretation" in which "a particular set of characters and actors," take on particular ascribed meanings and ways of being (Holland et al., 1998, p. 52), such as Chavo, the bumbling orphan in *El Chavo del Ocho*, or the precocious Pepito, who routinely encounters Fidel Castro in Havana. A figured world is not only a physical space, such as the *vecindad* in which Chavo lives, or the classroom where Pepito engages in mischief, but also a culturally mediated environment in which social norms and discourses explicitly and implicitly shape the identities constructed by its "actors." As identities are fluid, and constructed in the context of an individual's interactions within a figured world, the cultures that mediate figured worlds are also dynamic and continuously changing.

As I have argued elsewhere, the notion of cultures in transition evokes the multiple ways in which cultures—as well as the identities that they recursively shape and inform—interact with factors such as socioeconomic class, politics, and citizenship (Caraballo, 2012). For example, current debates about citizenship in Latin America incorporate issues ranging from the politics of identity to the connection between cultural diversity, identity, and citizenship in the experiences of Latina/o migrants, in which the processes of globalization are often invoked as a way to conceptualize changing ideas about citizenship (Dagnino, 2005). However, as Guerra argues, for people of color whose communities are impacted by immigration, globalized notions of identity overlook the reality and significance of the continuous crossing of figurative and literal borders in immigrant communities. Guerra proposes the concept of *transcultural citizenship* as one that "possesses more explanatory power than global citizenship … for understanding our commitment to both global *and* local" identities and cultural contexts (2016, p. 85).

As such, a figured world exists in relation to other cultural boundaries and contexts, such as those determined by individuals' local, national, or ethnic affiliations, which in turn shape what counts as humor within and beyond cultural groups. Chespirito's comedy is iconic in its ability to supersede national and regional borders, leading to its association with a broader Latin American identity. However, the idea of a Latin American identity can be problematic when it obscures the situated nature of our experiences, including our experiences of humor. As Alharthi argues, humor is a social phenomenon, and while many communities "may share the same concept of what humour is," as many Latin Americans have enjoyed Chespirito's sketches for decades, they also have "unique techniques . . . for provoking laughter which, in many instances, distinguish them from others" (2014, p. 4). Arguably, my friends, family, and I enjoyed Chavo's antics and Pepito jokes as part of a particular figured world, one that was discursively produced by the social and political contexts in which we experienced them as exiled Cubans in Puerto Rico or in the United States. Thus, in this chapter I explore how thinking across cultural and national borders (among others) can help shape transcultural conceptions of citizenship and identities that attend to both the local and the global in the Latin American figured world where audiences experience the humor of Chavo and Pepito.

Chespirito and Pepito as cultural icons

The creativity and humor of Roberto Gómez Bolaños, renowned as *Chespirito* after well-known director Agustin P. Delgado called him a "Little Shakespeare" (Gómez Bolaños, 2006), led to the great success of his sketches all over Spanish-speaking Latin America and Brazil, United States, and Spain (Rodríguez Valle, 2014) and their subsequent translation into European and Asian languages (Feng et al., 2014). While some argue that the sketches were originally written for adults or older children (Feng et al., 2014), Chespirito's series and their main characters were popular with children and generally considered ideal (and at times educative) programming for the whole family, given its implied lessons about social values, behavior, and language (Luque Duran, 2007). Their reception, in turn, is connected to particular desires and hopes inscribed in the identities and figured worlds where the humor was experienced.

Chespirito: Entertainment and social commentary across Latin America

Beyond a number of published works written by or about Chespirito himself (e.g. Gómez Bolaños, 2005, 2006, 2013), the sheer volume of entries and reviews about Chespirito's series and sketches on the popular Internet Movie Database (IMDb) site is one indicator of their reception within and beyond Latin American publics. Although sites such as the IMDb are open forums in which anyone can add comments or edit information, and therefore lack the veracity and trustworthiness of professional reviews and refereed sources, they offer a basic sense of a broad public's range of impressions and responses to the shows. As of this writing, an IMDb entry for *El Chavo del Ocho* (1972) describes the series as "the misadventures of a poor, homeless and fumbling boy and his friends in the humble neighborhood they live in," with thousands of reviews from audiences all over the world that profess the general appeal of the series, particularly for families.[2] In addition, widely available reviews on social media platforms from scores of viewers and fans, and approved archival sites such as www.chespirito.com and www.chavodel8.com, immortalize the sketches and their characters. First produced in 1971, on the air in various formats until 1992, and since then available in syndication, the show is still one of the most lucrative franchises of the Televisa network (Lopez, 2014).

However, while critics and scholars have drawn from the widely popular creations of Gómez Bolaños to examine issues in sociology, linguistics, and psychology,[3] few if any, of these popular reviews or critiques and scholarly analyses explore the potential for political commentary in Chespirito's sketches, whether explicit or implicit, nor how its appeal to broader Latin American audiences might have implications for notions about identities and citizenship in transnational contexts. One exception is a feminist/postcolonial reading of male tropes such as the *pelado*[4] in Mexican cinema (Quiñones, 2014). In her analysis of *pelados* in Mexican popular culture, Quiñones seeks to "expand the historical understanding of these Mexican comedic performers" and determine whether their texts catalyze or stunt social progress and hope by exploring the *pelado* as a key figure among "society's undesirables—la *chusma*" (p. 12). According to Quiñones, Chespirito's *pelados* are "constructed via shame and obligation in poverty, offering no improvisations or socio-political occasions for change" (p. 13).

Regardless of their apparent complacency in their socioeconomic contexts, the characters' interactions with others are situated in their environments and cultural worlds, leading to divergent, even if overlapping, narratives of poverty, agency, and hope. As discussed below, Pepito's sharp quips and bold exchanges with political leaders are also grounded in the context in which these jokes are told, whether on the island or among Cuban communities in exile. Each of these contexts reflects cultures in transition across generations, socioeconomic boundaries, and national borders. Thus, given their very different origins and purposes, yet similarly ubiquitous popularity, juxtaposing Chavo with Pepito offers an opportunity to examine how humor, in the context of cultures in transition, can inform the identities and experiences of the participants in this narrative inquiry, who them/ourselves cross national, generational, and cultural borders.

Pepito: Humor and sarcasm in revolution and exile

As an eternal child figure in many Latin American and Spanish-speaking cultures, tales of Pepito (and his namesakes elsewhere) serve as a common point of reference among different cultures and nationalities. While often associated with jokes in "poor taste" or double entendre, there are also those that are charmingly reminiscent of childlike innocence and draw from traditions of physical comedy similar to that of *El Chavo del Ocho*. For example, one day:

> *Pepito le dice a su mama: Yo ya no quiero jugar más con Juanito al rompecabezas. ¿Por qué? Porque al primer martillazo ya está llorando.*
>
> [Pepito tells his mother, "I don't not want to build puzzles with Juanito anymore." "Why not?" "Because he starts to cry at the first attempt."][5]

In Cuba, Pepito jokes range from the escapades of a mischievous boy, similar to the joke above, to those which convey a more pointed narrative of sociopolitical critique. According to Yoani Sanchez, the Cuban dissenter and journalist who was named among the 100 most influential people of the year by *Time Magazine* in 2008:

> Pepito, our eternal rogue child, has been to the moon, to hell, to the Vatican and has, on several occasions, crossed the straits of Florida. From his guilt-innocence he has suggested irreverent solutions and on more than one occasion he has been more lucid than the analysts and the academics (2007, n.p.).

The Cuban Pepito, whether directly or indirectly, often voices polarizing economic, social, and political sentiments that Cubans on the island, given its restrictive political regime and censorship, might not dare utter. These Pepito jokes, in particular, offer a respite during difficult circumstances, and a release valve for independent thinking in oppressive times.

Also in contrast to Chespirito's sketches, in which the political is most often implicit, many (though not all) of the jokes featuring Pepito are overtly political, due to the character's organic origins. Pepito is a social construct, one that reflects the cultural and political discourse of his time and context, a post-revolutionary Cuba from the 1950s to the present. In Cuba, where "socially, sarcasm is Cuba's predominant psychological staple for coping" (Abril, 2013, p. 14), humor is commonly political despite great (and often successful) efforts at censorship. In an internationally recognized blog, *Generación Y,* within the broader platform of *14 y Medio*, founded by Sanchez and maintained by dissenters on the island, a recent post reported on current attitudes toward the mandatory slogans of the *pioneros* [pioneers], as the government-run education system calls the schoolchildren. In this post, the author illustrates current attitudes about the slogan, and the educational doctrine it encapsulates, with the following joke, where Pepito:

> *llega a la escuela y le comunica a su maestra que la mascota de su casa ha parido diez gaticos. Unos días después, la solícita profesora indaga sobre la salud de la camada y recibe una inusual respuesta. Cinco gaticos ya abrieron los ojos y los otros siguen siendo comunistas, apunta con agudeza el chiquillo* (Sanchez, 2015, n.p.).
>
> [Pepito comes to school and tells his teacher that his pet cat has given birth to ten kittens. A few days later, the solicitous professor asks about the health of the litter and gets an unusual answer: "Five kittens have opened their eyes, and the others remain communists."]

In the same manner, Pepito makes regular appearances in the most politically charged texts, such as Sanchez's commentary in anticipation of US President Obama's visit to Cuba in March 2016:

> Pepito, the little boy who stars in our popular humor, will release a couple of jokes for the occasion, and tchotchkes sellers will offer items with the lawyer's profile ... [The President's words,] however, ... should be directed to those young people who right now are assembling a raft, fueled by their despair they carry within (n.p.).[6]

In referring to the new jokes that will be "released" for the President's visit and the new products that emerging entrepreneurs will peddle, Sanchez is also implicitly addressing the generativity of this humor, the creative agency that leads to the joke or the newly minted commemorative item for sale. While Pepito's character is continuously and collaboratively authored by those who tell his jokes, Chespirito's Chavo, among other characters, was his own creation. The connection between humor, identities, and cultures, broadly defined, is thus manifest in audiences' experiences of Pepito and Chavo, as exemplified in the narratives from the participants in this inquiry.

A narrative inquiry

Building upon the interconnectedness between humor, identities, and cultures in audiences' experiences of Chespirito's shows and Pepito's jokes, this inquiry examines themes that emerge in participants' narratives in the context of their own cultures in transition. Parallel to the interrelatedness of cultural discourses and identity construction, the process of constructing the self has been associated with narrative-making by scholars in various fields of research and scholarship (Clandinin et al., 2016; McCarthey and Moje, 2002; Ochs and Capps, 1996; Smith and Watson, 2000). One main point of consensus among a range of perspectives about narratives is that they are a fundamental part of a person's ability to make sense out of experiences. According to Ochs and Capps (1996), "narrative and self are inseparable in that narrative is simultaneously born out of experience and gives shape to experience" (p. 19). In this context, our experiences in community and with/in exile from our home countries are part of the meaning-making process in relation to being an audience for Chespirito's scripts and Pepito jokes. Like Bruner (2004), "the approach I shall take to narrative is a constructivist one—a view that takes as its central premise that 'world making' is the principal function of mind, whether in the sciences or in the arts" (p. 691).

In this chapter, I conceptualize narratives as tellings, writings, or depictions in which individuals articulate their ideas, experiences, and identities, conveying "versions of reality" and/or "embodiments" of points of view (Ochs and Capps, 1996, p. 20). These narratives, which I co-construct as part of my inquiry, are drawn from a variety of methods, ranging from conversations to written responses. In each circumstance, the prompts were presented in Spanish, and offered as a way to invite participants' recollections about episodes of *El Chavo* that they recalled and any of the Pepito jokes that they knew, as well as their thoughts about the cultural significance of these episodes and jokes. Due to constraints of time and distance, the narratives were collected in the form of written reflections from 12 friends and family members whom, when polled, remembered both watching Chespirito's programs and had either told Pepito jokes or heard them from family and friends who lived or live in the United States.

Although Portelli (2003) suggests that "oral sources are never anonymous nor impersonal, as written documents may often be" (p. 14), because these writings were created for this purpose, they convey an additional step in what Bruner describes as "processes of selection" (1971, p. 6) as each person re-constructed and recorded his/her memories and impressions, reflecting the choices made about which memories to share and/or which episodes or jokes to recount. I placed particular emphasis on the narratives co-constructed by the participants and myself, as well as their actual references to the subjects of Chespirito and Pepito, as we sought to make meaning of our experiences as immigrants from

Latin American countries and first-generation Latinos in the United States (Clandinin and Connelly, 2000; Connelly and Clandinin, 1990). While conceptualizing the cultures in transition of multiple generations of immigrants from various countries renders a Latin American identity problematic, there is arguably discursive power in the ascribed construct of being "Latino/a" in the United States. As I recorded notes on conversations with family and friends as participants, and reviewed the letters and anecdotes they sent me, I analyzed them via an iterative process of culling the themes that recurred across multiple narratives (Bogdan and Biklen, 2007).

Chespirito and Pepito across cultural and national borders

In this iterative and interpretive process (Denzin, 2002), the themes that emerged during my narrative analysis, drawn from very divergent origins and life experiences, underscore the deeply complex and wide variety of circumstances and contextual factors that impact the ways in which intergenerational and intercultural lessons and values are transmitted via family memories and stories. The sociopolitical context of our lives, as well as language traditions and culture, are among the many factors that facilitate the transmission of family memories, experiences, and values via the stories that are passed on to future generations and for those of us who are immigrants, across national and cultural borders. In this analysis, Leichter's (1974) observations about the study of families are particularly relevant and provide important insights toward our understanding of the connections between individuals' experiences and their values, perspectives, and worldviews: "experiences at the margins of consciousness or at the level of peripheral awareness . . . remain part of the educative process" (p. 26). In the broadest sense, the educative process that Leichter alludes to is the constructive process that shapes participants' identities and cultural contexts. These implicit and explicit social processes become increasingly evident in the intergenerational nature of the narratives about Chespirito's sketches and Pepito jokes via themes that recur in participants' narratives: nostalgia, innocence, and goodwill; humor and politics; and social class and poverty. Below, I include excerpts from the narratives of four participants (introduced from oldest to youngest) that exemplify the themes above, albeit from individual and often divergent perspectives: Zoraida (my mother), who emigrated from Cuba to Puerto Rico in the early 1970s in her early twenties; Teresita, who emigrated from Cuba to Costa Rica with her husband and two sons in the 1980s; Chely, who permanently resides in Cuba but lives several months out of each year in the US; and Soribel, who was born (and lives) in New York City to a Dominican and Haitian family and maintains close ties to her family in Santo Domingo and Haiti.[7] The varied backgrounds and experiences of the participants convey some of the diversity among Latinos who live or spend significant amounts of time in the US. I focus on these themes because they were represented in the majority of the recollections I examined, whether oral or written, and all were grounded in notions of family-friendly intergenerational humor, messages about daily struggles, goodwill toward others, and hopes for the future.

Storying nostalgia, innocence, and goodwill

When I began to conceptualize this chapter, I asked several colleagues, friends, and family members whether they would be interested in talking with me and with others about their Chespirito and Pepito stories. Over the course of several months, I received texts, emails, and handwritten responses with jokes, stories, and reflections. My mother was probably the one who did the most fieldwork on the topic, so to speak, as she found it fascinating to come across so many friends and acquaintances from different backgrounds that had similarly fond memories of Chespirito, particularly Chavo. In recounting the various countries of origin and migration pathways of the people with whom she conversed about these shows, she also conveys her own nostalgia for Chespirito's sketches. She came to the conclusion that others remember the show so well because of its comic appeal, combined with the wholesomeness of the comedy, sans violence[8] and profanity, and laments that these programs could not be enjoyed by families in Cuba. According to my mother:

> *Supe que Maite y sus hijas cuando salieron de Cuba fueron a vivir a México por muchos años y gozaron del programa de* El Chavo del Ocho, *y* El Chapulín Colorado, *... pues entretenían a toda la familia de una forma amena y con libretos tan reales como la vida misma, sin palabras obscenas, ni armas, ni violencias. También en Colombia, una terapista llamada Ana vivía allá, pues ese era su país de origen, hasta que vino para Estados Unidos, y me habló muy bien de esos programas ya que también sus hijas esperaban esa hora con ansias. En Costa Rica, se veía también, y es posible que el único país latino donde no pudieron los niños gozar de ese programa fue en Cuba por no tener el sistema de comunicaciones que en el resto de la América Latina. Que días tan lejos ...* (Zoraida)

> [I knew that when Maite and her daughters left Cuba to live in Mexico they enjoyed the programs of *El Chavo del Ocho* and *El Chapulín Colorado* . . . the shows entertained the whole family in a pleasant way, with lifelike scripts without obscene words, weapons, or violence. Also in Colombia, a therapist named Ana who lived there because it was her country of origin, until she came to the United States—spoke highly of those programs, and mentioned her daughters also anxiously waited for the show to come on. In Costa Rica, it was also seen, and it is possible that the only Latin country where children could not enjoy the program was in Cuba because it did not have the communications system that the rest of Latin America had. So long ago . . .] (Zoraida).[9]

Her own recollections also teemed with nostalgia as she recounted how sharply at seven each evening, our whole family in Puerto Rico watched *El Show de Chespirito* without fail, concluding her vignette by musing: *que días tan lejos*, so long ago. And, as I can personally attest, my family being a fairly sentimental bunch, we often shed a tear or two when someone was mean to Chavo, who had a generous heart and cared for his neighbors, despite managing all on his own and with so few possessions:

> *No nos perdíamos un capítulo diario y algunas veces llorábamos pensando que El Chavo era de verdad un niño pobre y sin familia pues vivíamos muy seriamente el personaje. Recuerdo un programa triste cuando por error pensaron en la vecindad que El Chavo había robado algo y él muy triste se iba con sus cositas colgando en una vara, y al darse cuenta todos los vecinos salieron y se lo impidieron, pidiéndole perdón (*Zoraida*).*

> We would never miss an episode and sometimes we would cry thinking that El Chavo was really a poor child with no family because we lived that character so vividly. I remember a sad episode when they mistakenly thought that El Chavo had stolen something and El Chavo, very sadly, begins to leave with his items hanging from a stick, and when the mistake is realized, all the neighbors come out and stop him, asking for forgiveness (Zoraida).

I remember watching that episode with my parents, when I was very young, exasperated by how obvious the thief was (Sr. Hurtado—"*hurtar*" means to steal in Spanish), angry at how unjustly El Chavo was treated, and heartbroken at his disappointment when others turn on him. Yet of greatest interest to me now is how that episode conveyed what I understood to be the message of the sketch. After leaving the *vecindad*, Chavo visits a priest who tells him to return if he has a clean conscience, and asks him to pray for a resolution to the situation. Quico asks Chavo if he prayed for the thief to get caught, and Chavo replies no, that he prayed for the thief to repent, because the most important thing was for him also to have a clean conscience. The thief overhears, and all is resolved when he returns all belongings and gives Chavo one of his beloved *tortas de jamón* [ham sandwiches]. Chavo, in turn, despite being perpetually hungry, shares his torta with Quico. It is precisely this perception of his innocent goodwill that many reminisce about when they think about the sketches:

> *Todavía mayores, todavía vemos las grabaciones en los canales de cable y nos da la misma emoción y también la risa pues algunas veces es un poco desesperante con sus torpezas* (Zoraida).
>
> Even as adults, we still see some old episodes on cable and it gives us the same excitement and that same laughter, because sometimes he is so over-the-top with his clumsiness (Zoraida).

Beyond the messages often conveyed by the sketches, there were also the technical effects, physical comedy, and playful scripts that appealed to families, including children:

> My earliest memories of Chespirito were in the early–mid-80's. My mother's family would always travel back and forth from the DR [Dominican Republic] and would always make references about what Chespirito or El Chavo would say ... there were some animated movements or sounds that were associated with the animation that would make us laugh (Soribel).

The intergenerational appeal of the comedy was achieved by the animation and other technical effects, which were unusual in live action programming at the time (Gómez Bolaños, 2006), and made the show appealing to children, even when they did not understand all of the jokes. The antics of the characters were also generally appealing to communities in various parts of Latin America:

> Culturally, Chespirito has been a common thread to all Latinos ... [M]any of us in high school would joke around with one another and would mock Don Ramón's "!Ya, ya, ya ...!" when we were exhausted after state exams. It was certainly an inside joke among us no matter what ethnicity (Soribel).

A similar nostalgia about the shared experiences of humor among her cousins is evoked by jokes and tales about Pepito:

> There is a sense of nostalgia that brings us together when Pepito comes up ... Anytime we heard simple words that were used to describe things that we could not say as kids, we would laugh inconsolably. The anticipation of the punchline alone made us laugh—even if we did not understand the entire joke (Soribel).

Like Soribel's experience with the Chespirito shows and Pepito jokes that she enjoyed even if she did not understand all of the content due to its language, its familiarity and physical humor led to a shared delight in the punchline. As suggested by the narratives above, despite the diverse contexts for these experiences, spanning from the US to Puerto Rico and the Dominican Republic, nostalgic accounts of Chespirito and Pepito are associated primarily with feelings of innocence, goodwill, and nostalgia experienced and shared by their audience decades prior.

Word play and politics

Despite similarities in the nostalgic reminiscences inspired by both characters, there are significant differences in their engagement with the political context in which Chespirito's sketches aired on television and Pepito jokes are told. According to Stavans,[10] "the content of the [Chespirito variety] show is utterly apolitical" (2014, n.p.). Basing his assertion on the popular sense that Televisa, "the Mexican media conglomerate [that is] often seen as a mouthpiece of the Institutional Revolutionary Party (PRI), endorsed the show without reservations throughout its duration," Stavans notes that the show's "repertoire overflowed with stereotypes like the crying girl, the fat real estate developer and the goofy teacher." While the use of these tropes would support Stavans' argument about the kind of entertainment that would be endorsed by the Mexican national media conglomerate, I would argue that rather than being apolitical, the sociopolitical commentary in Chespirito's comedy is implicit, particularly with respect to social class and poverty, as discussed in the next section, while Pepito's political agenda is often much more explicit.

In contrast to Chespirito's perceived focus on moral and social values, Pepito jokes and folktales are often centered on double entendre and word play, as exemplified in a very popular joke below that is regularly circulated in social media and was present in more than half of the narratives from participants. In this joke, Pepito innocently reveals a family crisis by using the word "revolución"[11] to describe a chaotic family situation, while simultaneously delivering a political pun:

Pepito llega a la escuela, y dice la maestra de español, Hoy vamos a dar la clase de los sinónimos. ¿Cuál es el sinónimo de la palabra regla?
Dice Pepito, Revolución.
Responde la maestra, Revolución no es sinónimo de la palabra regla.

> *Pepito le responde, ¿Que no, Maestra? Si a mi hermana hace tres meses que le falta la regla, y en mi casa hay una revolución tremenda* (told by Chely).
>
> [Pepito arrives at the school, and the Spanish teacher says, "Today we are going to study synonyms. What is a synonym for the word ruler?"
> Pepito says, "Revolution."
> The teacher responds, "Revolution is not synonymous with ruler."
> Pepito responds, "No, Miss? But it has been three months since my sister had her 'ruler',[12] and in my house there is now a tremendous revolution"] (told by Chely).

Beyond references to national politics, like Cubans on and off the island, Pepito comments on the international context of Cuban politics:

> *Llega Pepito al aeropuerto de la Habana. Entrando a la aduana un grupo de cubanos de la isla lo reconocen y le dicen: ¡Pepito, el de los cuentos! ¡Caballero, que tiempo hacía que no venías por acá! Y Pepito le responde, ustedes se equivocaron, yo soy Pepito el de los chistes—los de los cuentos son Raúl y Obama.*
>
> [Pepito arrives at the airport in Havana. Entering through Customs, a group of Cubans recognize him and say, "Pepito, the one from the jokes! Sir, it has been so long since you were last here!" To which Pepito responds, "You are mistaken, I am still Pepito, the joker—the ones who tell stories are Raul and Obama."]

With their political reality as part of the audience's trajectory, it became intertwined with experiences of shows like Chespirito's despite its evasion of such themes. For example, Teresita explains why *El Chavo el Ocho* was so appealing to her family:

> *Cuando nosotros llegamos de Cuba en 1981, a Costa Rica, el programa que más nos gustó en la televisión fue el de* El Chavo del Ocho. *Era muy ameno y familiar, en todos los sentidos ya que nos hacía reír a nosotros dos y a los muchachos, pues en Cuba en aquel entonces lo que se veía en la televisión era los muñequitos rusos . . . la Revolución eliminó todo lo que olía a Estados Unidos* (Teresita).
>
> [When we arrived to Costa Rica from Cuba in 1981, the television program we liked the most was *El Chavo del Ocho*. It was very pleasant and familiar in every possible way because it made us and the kids laugh, especially in those days, when all you could see on Cuban television were Russian cartoons . . . the Revolution eliminated everything that hinted of the United States] (Teresita).

Teresita's recollection is specifically situated in the surveillance policies of 1970s Cuba, in which programming for children was generally Russian, and anything Western was automatically suspect. Although *El Chavo del Ocho* had no known political affiliations, in a country in which media is strictly managed and censored, programming came only from nations that were allies of the Cuban government. Beyond its primarily family-friendly themes, being able to watch a greater variety of shows was a welcome respite from the restrictions under which Cubans on the island had been living.

Social class and poverty

Although Chespirito's sketches are less overtly political than the Cuban jokes featuring Pepito, there is implicit social commentary about class and poverty in Mexico that resonated with viewers elsewhere. Much of the comedy in *El Chavo del Ocho* stems from what was described as the life of the urban poor in the *vecindad*—or, as it is known in Cuba, *el solar:*

> *El tema del programa era el solar, como le decían en Cuba a las casas de los interiores, o sea que no daban frente a la calle, y en México le dicen vecindad. Sin lugar a dudas, estos tipos de programas son muy buenos, pues tratan de vida cotidiana de la gente de a pie, o sea, la clase trabajadora* (Teresita).
>
> The theme of the program was the *solar*, what they call the inward-facing houses in Cuba, that is, they faced a courtyard rather than the street, and in Mexico they call it *vecindad*. Undoubtedly, these types of programs are very good because they deal with the everyday life of ordinary people, that is, the working class (Teresita).

At a time when many popular comedies were centered on the lives of suburban and middle-class families, audiences everywhere could relate to a situational comedy that focused on concerns that were not too removed from their own lives, or from the issues of their neighboring communities, such as the real-life implications of poverty, hunger, and the scarcity of employment. Don Ramón, who is always trying to make ends meet by taking up odd jobs and pawning his few belongings, is perpetually late with his rent. When the landlord insults his honor, he exclaims: *¡Ay Señor Barriga! Es que cuando el hambre aprieta, la vergüenza afloja. ¿Sabe usted lo que traigo desde anoche en el estómago?* [Oh, Mr. Barriga, when one's stomach is empty, it is harder to feel embarrassment. Do you know what is the only thing in my stomach since last night?]. While he may be too poor to be shamed by having to evade the landlord when he comes to collect the rent, as the only unemployed adult male in the show (and also a single father), he is often combative with others who imply that he is lazy.

In her comments about the *vecindad*, Teresita draws parallels to life in the Cuban "*solar*", adding her perspective on a portrayal of the experiences of the working class, as its members are ascribed and construct classed identities. Indeed, in *El Chavo de Ocho*, the *vecindad* was the common point of reference for all characters, even if they came together for different reasons. The Mexican *vecindad*, like urban tenements elsewhere, was the focus of controversial depictions of what was construed as a "culture of poverty" by anthropologists and policymakers who found cultural connections among impoverished communities in various parts of Mexico and Latin America (Lewis, 1975). The culture of poverty paradigm made no explicit connections to the structural and political conditions that historically led to poverty and unemployment, which worsened during the economic crisis of the 1980s as Mexico adopted increasingly neoliberal policies (Latapí and González, 1995). It is perhaps the very relatable portrayal of the *vecindad*, at the center of the lives of his memorable characters, that most concretely exemplifies both the universal success of Chespirito's sketches (exemplified in viewers' ability to relate them to their own lives, as in Teresita's comment above) and their simultaneous avoidance of potentially polarizing political stances in Mexico. While Chavo's poverty is fully embodied in his memorable character, it is not situated in Mexico's particular socioeconomic and political realities,[13] and its portrayal of poverty and class remains ripe for further interpretation.

Doña Florinda, for example, who came from greater socioeconomic means, became widowed and moved to the *vecindad* when she had no other recourse. Most of her dialogue in the show is devoted to distinguishing herself and her son, Quico, from the "low class" *chusma*. As an example, when Doña Florinda commissions Don Ramón to make a sign for the courtyard to prohibit rowdy play in the common space, even though it is riddled with misspellings, she hangs it up and says, "*Para la vecindad esta bien, para que lo entienda la chusma.*" [For this neighborhood it will be fine, the rabble will understand it]. While Doña Florinda is often ridiculed for putting on airs, the overall premise of *El Chavo del Ocho* avoids explicit commentary about the political and economic factors that exacerbated poverty and unemployment in Mexican cities in the 1970s and 1980s (Fleck and Sorrentino, 1994) which underlie the social conditions portrayed by its sketches.

Pepito jokes, on the other hand, frequently bring political figures and Cuba's political context into conversation, making direct links to the scarcity of food and resources on the island:

> *Una vez iba caminando Pepito, y se tropieza con el Comandante Fidel Castro que estaba recostado a un árbol muy triste y preocupado. Pepito le pregunta,*

> *"Comandante, ¿qué le pasa?" Y él le responde, "Aquí, Pepito, estoy abatido." A lo que Pepito le responde, "Menos mal, Comandante, que usted está a batido, pues nosotros sólo estamos a pan y agua"* (told by Chely).
>
> [One day, Pepito was walking along when he bumped into the Commander-in-Chief, Fidel Castro, who was leaning against a tree looking sad and worried. Pepito asked him, "Commander, what is wrong?" And he responded, "Pepito, I feel shaken." And Pepito responded, "At least you are better off than we are. We just have bread and water."][14]

Pepito's astute retort is a commentary on the reality of many Cubans, who find themselves making do with the most limited provisions provided by the government. His reply also exposes the hierarchies present even in a "classless" society, implying that Castro, even when he is not doing well (as living off milkshakes might suggest), is still better off than the general public who are surviving on bread and water alone.

Although Pepito's jokes, centered on an oral figure whose comedy must be mentally visualized by audiences, are less concrete in their portrayal of complex issues about socioeconomic class and politics, they emanate from the complex perspectives about politics, class, and economics of Cubans on the island and in exile. Pepito's jokes are directly situated in the nuanced context of his home country's communist regime, where all sacrifices are justified as necessary for the common good. Thus, while the character of Pepito is an eternal child figure, similar to Chavo, the Cuban version of Pepito is continuously (re)-constructed as a product of the socioeconomic and political concerns and commitments of its multiple authors, as another author contributes to Pepito's persona every time a new joke is told.

For three of the four participants whose narratives are included in this chapter, who shared a Cuban background, sharing Pepito's jokes generated conversations about the economic and political state of affairs in Cuba. In contrast, recalling favorite Chavo episodes led to discussions of family values, childhood memories, and/or a general sense of nostalgia about a shared pleasure. For Soribel, whose background is Dominican and Haitian, her reminiscences about both figures were more similar to each other. In all circumstances, humor led to shared experiences among family members and friends; these experiences were inherently situated in the dynamic cultural contexts negotiated by the participants. Thus, whether or not the political role of the character was more or less explicit, participants' experiences of humor and nostalgia demonstrate the interrelatedness of identities and cultures in the diverse intergenerational publics of Chespirito's sketches and Pepito's jokes.

Y ahora, ¿quién podrá defenderme? ["And now, who will help me?"][15]: Border-crossing possibilities

Several decades after watching *El Chavo del Ocho* as a child in Puerto Rico, and having lived in various parts of the United States, I have been struck by Chespirito's continued popularity, particularly after his death in 2014. His unforgettable characters exemplify the significance of his comedy across Latin American nations, prompting viewers to consider what has led to the prolonged resonance of his iconic characters within and across such diverse cultures and communities. As Stavans mused, what united Latin Americans "was neither geography nor history but popular culture" and "Chespirito is no longer only Mexican," but a symbol of a broader cultural identity constructed out of the shared experience of his humor (2014, n.p.). Similarly, the Cuban version of Pepito is very much a national figure authored by the individuals who feature him in their jokes and stories, but carved out of a transcultural character recognized and appropriated by diverse Spanish-speakers. The border-crossing generative nature of these iconic figures can be sites of possibility for the construction of transcultural identities as Latin Americans continue to grapple with evolving and dynamic conceptions of citizenship in their home countries (Dagnino, 2005) as well as the impact of migration and immigration on cultural, national, and transnational identities (Flores, 1995; Smith, 2006).

According to Clandinin and colleagues (2016), "as narrative inquirers, we know that our work is to research, to try to understand, to systematically inquire into the phenomenon of experience—that is, the storied experiences of people" (p. 14). While there is cause for caution in assuming that the self is fully knowable or accessible via the narratives that individuals themselves construct (Miller, 2005), Clandinin and colleagues' call to narrative as inquiry highlights possibilities for building upon narratives, such as the reflections on experiences of Chavo and Pepito examined above. These narratives take on a catalytic nature as the inquiry offers opportunities to explore the multiple "borders" that we continually cross between identities and experience, the national and the local, freedom and repression, virtue and indifference, hope and despair, as conveyed in the participants' reflections. These border crossings have significant implications for the construction of our identities as agents of change in our society, and as citizens in our communities. According to Guerra:

> what we think of as citizenship must be seen as a direct consequence of the different ways in which language and culture are implicated in the production of

> a particular kind of identity, one that is fluid and multi-faceted but simultaneously acknowledges and responds to the ever-present linguistic, social, cultural and political opportunities and constraints that govern our lives (2015, p. 97).

As such, encounters with language and culture, via popular entertainment, family anecdotes, and interactions with media are rife with opportunities for agency in constructing identities as transcultural citizens, not just consumption of existing narratives of nationality and citizenship as Latin Americans. Rethinking citizenship in this way, as a consequence of the production of identities, rather than the other way around, also challenges us to think critically about public discourses such as those around globalization and pluralism. As Guerra contends, "we need to move beyond the inherent constraints that the term global citizenship forces on us all in its unintended elision of the local" (2012, p. 85).

In the same vein, Chespirito's humor arguably resonates with such notions of transcultural citizenship and identities, as viewers across generations and nations share similar experiences with the programs in my narrative inquiry group. Yet these recollections also convey the situated nature of our interactions with Chespirito's characters and Pepito's jokes, for example, as they are grounded in their own local social and political contexts. For example, foreshadowing a more public renaissance of agency and participation in a country where citizenship is continuously being redefined, Pepito is making a comeback in Cuba, and the appeal of Chespirito's sketches is ripe for discussion wherever his sketches are well known. Despite Sanchez's (2016) observation, almost ten years before, about Pepito's gradual retreat into the sidelines, on the recent occasion of President Obama's visit, many gladly welcomed him back to center stage:

> This morning, for a few hours, people will put aside conversations about high food prices and complaints about the collapse of transportation, aggravated by the security measures that plague the city. On the streets there is a resurgence of jokes starring Pepito, the mischievous child of our folktales, who emerged from his long silence to laugh even about the great visitor's mother-in-law (n.p.).

Humor is thus powerful, persistent, and enduring, even when it lies dormant for a time, as in Pepito's Cuba, or more universally venerated, as might be the case with Chespirito's sketches. Where there is humor, there is hope. Fostering inquiry in the context of humor; understanding the identities that we construct in our particular cultural figured worlds; and understanding citizenship as transcultural, is uncertain work, but it is also life-affirming, inclusive, and supportive of change and growth. Exploring the similarities and differences among audiences' recollections of figures such as Chavo and Pepito, whose appeal crosses many cultural and national borders, invites a deliberate engagement with some of the political and socioeconomic narratives that inform current social discourses and power relations in Latin American societies. Within and beyond local contexts, this kind of examination catalyzes the "conscious, practical work" (Freire and Macedo, 1987) of understanding the culturally grounded narratives that shape the past and present as we strive to imagine and shape our futures.

Notes

1 Pepito may have derived from Luciano Bottaro's eponymous comic book series, although I have not found an explicit connection between the Italian version of this character and the Cuban version of Pepito.

2 Reviews are available at: www.imdb.com/title/tt0222541/reviews?ref_=tt_urv and http://www.imdb.com/title/tt0229888/?ref_=nmbio_mbio.

3 Topics in these analyses range from an informal consideration of "superhero" personalities (Saldaña, n.d.); to a study of the metacognition of linguistic patterns in Mexican society (Luque Duran, 2007); to the effect of the absence of Spanish television programming on bilingualism (Feng et al., 2014); to the show's lack of appropriate behavioral role models for young girls (interestingly focusing only on girls; Rodríguez Valle, 2014).

4 According to Quiñones (2014), "the pelados of Mexico City were the peripheral proletariat regarded only in terms of labor and production. They were riff-raff" (p. 12).

5 In Spanish, a puzzle is a *rompecabezas*, a compound word make up of "breaks" and "heads." Since there is no similar word in English, the joke is lost in translation.

6 Many entries in *Generacion Y*, as well as its partner blogs and sites within the broader platform of *14 y Medio* are available in translation. The entries have been translated and updated, even during government blockages of the writers' access and shutdowns of their sites, by a network of contributors outside of Cuba. Sanchez regularly emailed her posts by passing as a tourist and purchasing expensive online access in Havana's Internet cafes.

7 Although the participants have a range of backgrounds, at the time of this writing, they all live in the US all or part of the time.

8 This observation is somewhat problematic, given that much of the physical "comedy" in *El Chavo del Ocho* involves hitting, punching, and slapping. In this context, however, my mother frames "violence" as that which leads to death and bloodshed, as in action genres.

9 Unlike other Latin American countries, who had access to similar international television channels and radio transmissions, Cuban television and radio are strictly censored by the government, which limits what programming is available on the island.

10 Ilan Stavans, a professor of Latin American and Latino culture at Amherst College, MA, noted that his father was a guest actor in the *Show de Chespirito.*

11 An allusion to the Cuban Revolution, during which Fidel Castro came to power in 1959.

12 In Spanish, *regla* also refers to a woman's menstrual cycle.

13 For example, although Mexico's unemployment rates are typically high, Don Ramón's recurring lack of income reflects an ongoing debate about how unemployment rates in Mexico are calculated, which counts as employed even those who subsist on small jobs that do not provide a stable income (Fleck and Sorrentino, 1994).

14 This joke contains a play on the words "*abatido*" [shaken up] and "*a batido*" [a milkshake]. Like many other plays on words, it is lost in translation.

15 This is the phrase uttered by every "victim" that Chapulín Colorado eventually rescues.

References

Abril, R. (2013). *Pepito's Cuba and the avocado baseballs.* Self-published. CreateSpace Independent Publishing.

Alharthi, A. (2014). "Humour and Culture." *International Journal of Humanities and Cultural Studies, 1*(3).

Bogdan, R. C., and Biklen, S. K. (2007). *Qualitative research for education: An introduction to theory and methods.* Boston, MA: Pearson.

Bruner, J. (1971). *On knowing: Essays for the left hand.* Cambridge, MA: Harvard University Press.

Bruner, J.S. (2004/1987). "Life as Narrative." *Social Research, 71*(3), 691–710.

Caraballo, L. (2011). "Theorizing Identities in a 'Just(Ly)' Contested Terrain: Practice Theories of Identity Amid Critical-Poststructural Debates on Curriculum and Achievement." *Journal of Curriculum and Pedagogy, 8*(2), 155–177. doi: 10.1080/15505170.2011.624939.

Caraballo, L. (2012). "Cultures in Transition." In J. A. Banks (Ed.), *The encyclopedia of diversity in education* (pp. 381–384). Thousand Oaks, CA: Sage Publications.

Cremin, L.A. (1974). "The Family as Educator: Some Comments on the Recent Historiography." *Teachers College Record, 76*, 250–265.

Clandinin, D.J., and Connelly, F.M. (2000). *Narrative inquiry: Experience and story in qualitative research.* San Francisco: Jossey Bass.

Clandinin, D.J., Caine, V., Lessard, S., and Huber, J. (2016). *Engaging in narrative inquiries with children and youth.* New York: Routledge.

Connelly, F.M., and Clandinin, J.D. (1990). "Stories of Experience and Narrative Inquiry." *Educational Researcher, 36*(5), 2–14.

Dagnino, E. (2005). *Meanings of Citizenship in Latin America. Working Papers* (Vol. 258). Sussex, England: Institute of Development Studies.

Denzin, N. (2002). *Interpretive interactionism* (2nd edn). London: Sage.

Feng, Y., Olague, R., Diaz, A., and Nzai, V.E. (2014). "Spanish Television Hinders Hispanic Children's Bilingualism Development in Texas." *Journal of Modern Education Review, 4*(6), 405–417.

Fleck, S. and Sorrentino, C. (1994). "Employment and Unemployment in the Mexican Labor Force." *Monthly Labor Review*, 3–31.

Flores, J. (1996). "Puerto Rican and Latino Culture at the Crossroads." In G. Haslip-Viera and S. L. Baver, (Eds.) *Latinos in New York: Communities in transition*. Notre Dame: University of Notre Dame Press.

Gómez Bolaños, R. (2005). *El diario de el Chavo del Ocho*. Ciudad Mexico, Mexico: Punto de Lectura.

Gómez Bolaños, R. (2006). *Sin querer queriendo: Memorias*. Bogota, Colombia: Aguilar.

Gómez Bolaños, R. (2013). *Chespirito*. Madrid, Spain: Aguilar.

Guerra, J.C. (2012). "Cultivating Transcultural Citizenship in a Discursive Democracy." In C. Wilkey and N. Mauriello (Eds), *Texts of consequence: Composing social activism for the classroom and the community* (pp. 83–115). Cresskill, NJ: Hampton Press.

Guerra, J.C. (2015). *Language, culture, identity and citizenship in college classrooms and communities*. New York: Routledge.

Hall, S. (1996). "Introduction: Who Needs Identity?" In S. Hall and P. DuGay (Eds), *Questions of cultural identity* (pp. 1–17). London: Sage.

Holland, D., Lachicotte, Jr., W. L., Skinner, D., and Cain, C. (1998). *Identity and agency in cultural worlds*. Cambridge, MA: Harvard University Press.

Leichter, H. J. (1974). "Some Perspectives on the Family as Educator." In Leichter, H. J. (Ed.), *The family as educator* (pp. 1–43). New York: Teachers College Press.

Lopez, E.E. (2014). Roberto Gómez Bolaños, Mexico's comedic artist "Chespirito," dies at 85, *The New York Times*, November 28.

Luque Duran, J.D. (2005). "Los juegos lingüísticos: Fallos comunicacionales, humorismo verbal y reflexión metalingüística." In L. Luque Toro (Ed.), *Léxico Español Actual, Actas del I Congreso Internacional de Léxico Español Actual* (pp. 91–126). Venezia, Italia: Libreria Cafoscarina.

McCarthey, S., and Moje, E. (2002). "Identity Matters." *Reading Research Quarterly, 37*(2), 228–238.

Miller, J.L. (2005). *Sounds of silence breaking: Women, autobiography, curriculum*. New York: Peter Lang.

Ochs, E., and Capps, L. (1996). "Narrating the Self." *Annual Review in Anthropology, 25*, 19–43.

Portelli, A. (2003). *The order has been carried out.* Basingstoke: Palgrave MacMillan.

Quiñones, N. (2014). *The Pelado, Cantinflas and Chespirito: Humor and the negotiation of Mexican masculinities.* Unpublished dissertation, ProQuest LLC: UMI 3668675. Claremont Graduate University.

Rodríguez Valle, J.F. (2014). *Influencia de la serie comica televisiva "El Chavo del Ocho" en el comportamiento social de las ninas de primer ano de educacion basica del jardin de infantes "El Despertar" de la ciudad de Ambato.* Unpublished doctoral dissertation. Universidad Tecnica de Ambato. Ambato, Ecuador.

Saldaña, J.T. (n.d.). *Analizando al Chapulín Colorado.* Blog. Available at: www.sabersinfin.com/articulos–2/psicologia/670-analizando-al-chapul-colorado.pdf.

Sanchez, Y. (2007). Where's Pepito? *Generacion Y.* Available at: https://generacionyen.wordpress.com/2007/10/21/wheres-pepito/

Sanchez, Y. (2016). A visit more symbolic than political. *Generacion Y.* Available at: https://generacionyen.wordpress.com/2016/02/18/a-visit-more-symbolic-than-political/

Sanchez, Y. (2016). Obama is surrounded by symbols to win the hearts of Cubans. *Generacion Y.* Available at: https://generacionyen.wordpress.com/2016/03/20/obama-is-surrounded-by-symbols-to-win-the-hearts-of-cubans–14ymedio-yoani-sanchez/

Smith, R. C. (2006). *Mexican New York: Transnational lives of new immigrants.* Berkeley, CA: University of California.

Smith, S., and Watson, J. (2000). "Life Narrative: Definitions and Distinctions." In S. Smith and J. Watson (Eds), *Reading autobiography: A guide for interpreting life narratives* (pp. 1–48). Minneapolis: University of Minnesota Press.

Stavans, I. (2014). What "Chespirito" left us, Op-ed, *The New York Times*, December 4.

6

El Chavo del 8 as an "Intimate Public" in Venezuela

What Happened to the Good Life?

Erica Colmenares

Caracazo, El (February 27, 1989–March 1989): "The Caracas smash," a week-long wave of massive protests, riots, looting, and shootings that spread across Venezuela's capital and surrounding cities in response to the government's economic liberalization reforms (also termed neoliberal restructuring) and subsequent increase in public transportation and gasoline prices. Officials estimate that the military/security forces killed 276 people, with some reports estimating the number of deaths to be over 2,000. Massive political instability and social discontent followed.

* * *

I was only seven during El Caracazo, and like many other middle-class Venezuelan children, I was largely oblivious to the pandemonium outside my apartment building. The news outlets on television were reporting a "state of emergency" and "martial law," but I had no idea what those warnings meant. I only knew that school would be canceled for the week!

As the world outside unraveled, my twin sister, brother, and our caretaker, Berta, spent our days glued to the television screen, watching *El Chavo del 8* reruns, a Mexican comedy series about the adventures of a young orphan boy named El Chavo, and his colorful neighbors, all of whom live in a low-class housing complex. As devoted viewers of the show, we were irritated when the news would interrupt our programming to bring us disconcerting images from the world outside. But we were too enchanted with the show to care, too enmeshed in El Chavo's antics to worry about what was happening just three kilometers from our apartment.

Every once in a while, we would see the haggard, harried face of our father who—while running to and from his government job—would peek his head into our room to see how we were "holding up." *Holding up?* We were *feliz como una lombriz*! [happy like a worm].

Sometimes, my father would linger in the bedroom, glance at the screen and we'd laugh together at the hysterical adventures and mishaps that befell the show's characters, many of whom were children played by adult actors. But those moments of us laughing together were rare. We knew that the *something* happening outside was on our father's mind; his body, normally squishy and tender to our embrace, felt tense, rigid, and worried.

On day four of our *El Chavo del 8* marathon, during one of our father's visits where he sat down with us to watch the show, he quietly murmured something along the lines of, "So, this is what we have come to." Tears welled up in his eyes, but he quickly brushed them away before they made their way down his cheeks. "Ay, *papito lindo*," my sister and I cooed, imitating a phrase stolen from La Chilindrina, a clever girl who was often the mastermind behind the mischievously hilarious schemes. But, before we could hug him, *papito lindo* quickly stood up, turned off the TV, and barked, "Stop watching this garbage. No son *chusmas!*" [You are not rabble].

My twin sister, brother, and I looked at each other, puzzled and confused. *What was that about?* Berta stammered something along the lines of the show being "low class." *Huh*? We figured it probably had to do with the outside events that everyone kept talking about. But, because we really didn't understand that *something*, my siblings and I shrugged, closed the door, and turned the TV back on, continuing our *El Chavo del 8* marathon. We figured that's what El Chavo would have done, if only he had a TV.

* * *

Like those visceral experiences that linger on in the recesses of one's mind, only to be recalled years later, that moment—my inchoate sense that the outside world was "out of joint," my father's simultaneous disavowal and curiosity with *El Chavo del 8*, juxtaposed with my siblings' and caretaker's obsession with it—have stuck with me. And the show has "stuck" with millions of other Venezuelan viewers as well. While official statistics with respect to television ratings and audience demographics in Venezuela are hard to verify due to the country's poor accounting system in the field of communication (Bisbal and Nicodemo, 2006), a few empirical reports (in conjunction with my own anecdotal recollections) highlight *El Chavo del 8's* popularity during the Venezuela of my youth, the mid-1980s through 1990s. Broadcast on *Venevision*, the country's largest television network, during primetime, *El Chavo del 8* was cited as the program of choice by 61% of the population, a figure surpassed only by the nightly news and *telenovelas* (O'Sullivan-Ryan, 1996). Such figures raise a few questions: How does *El Chavo del 8*, a show originally marketed for children, become such a powerful site of emotional attachment among a group of people, including adults, in a country far from its intended audience? How does this series generate such powerful, varied, affective responses? And how did this "comedy show" gain such prominence in the Venezuela of my childhood (1980s to 1990s)—an era of political turmoil and "socioeconomic decay" (Lander, 2005)—punctured by events such as El Caracazo?

Guided by the theoretical possibilities afforded by the Affective Turn (Clough and Halley, 2007) which consider feelings, emotions, and affect as serious objects of scholarly inquiry (Cvetkovich, 2012, p. 3) and Lauren Berlant's (2008, 2011) extensive scholarship on affects, collective attachments, social belonging, and notions of the good life, I argue that *El Chavo del 8's* magnetism in Venezuela can be attributed to the show's capacity to forge an "intimate public" (Berlant, 2008): an aesthetic and affective structure that binds strangers together through emotional ties and promotes a feeling of belonging. Through the explicit and implicit chronicling of the waning of the fantasies of the good life (e.g., job security, upward mobility, social equality, durable intimacy, etc.) and its subsequent wearing out of the subject (Berlant, 2011), *El Chavo del 8* functioned as a space of emotional connection and reciprocity, sustaining viewers across different social classes as they discovered that their world (i.e., the Venezuela of the 1980s and 1990s), could no longer sustain the fantasies of a good life (Berlant, 2008; Cvetkovich, 2011). By offering multiple sites of identification for its diverse audience, the series was a sight/site of optimism and transcendence, and for some, experimentation with "better" forms of life. However, as I will argue, the show might have also been a site of *mis*recognition that was actually cruel in nature.

To do this, I begin with a brief discussion of the rise of economic liberalization policies in Latin America and then contextualize the specific precarious effects that neoliberal restructuring had in Venezuela during the 1980s and 1990s. Second, I describe how *El Chavo del 8*—through scenes of *crisis ordinariness*, or the ordinary dramas of adjustment to the "living within crises" (Berlant, 2011) of economic liberalization—helped to cultivate an intimate public and provide viewers with scenes they could both identify and recognize. In the latter part of the chapter, I explore how the show, and the intimate public it both forged and was forged by, may have been animated by a logic of "cruel optimism" (Berlant, 2011), a relational and affective double-bind in which attachment to particular "objects"—here, *El Chavo del 8* and notions of the good life—were both sustaining and diminishing (Berlant and Seitz, 2015). As an individual who formed part of this intimate public, this is a somber realization. Despite this hard-to-stomach recognition, I conclude with a description of how the series, and particularly the character of Don Ramón, might continue to offer viewers strategies for trudging through ordinary life. The significance of this chapter lies in exploring how mass-mediated "texts" such as *El Chavo del 8* can generate affective spaces for strangers to come together to dream, work, or try out alternative attachments, forms of adjustment, or modes of relationality in a world that continues to fray at the seams. And since I, too, formed part of this intimate public, throughout the chapter I include examples of my own attachment to the show because to speak of the general, it is necessary to speak of the personal. After all, since "the personal is the general" (Berlant, 2009, p. vii), personal narratives can be read as autobiographies of collective experience.

Rise of economic liberalization policies in Latin America and its accompanying precarity

El Chavo del 8, originally broadcast in Mexico between 1971–1980, reached Venezuelan television screens in the late 1970s, coinciding with the rise of economic liberalization that was immiserating much of Latin America. While neoliberal restructuring is not the subject of the chapter, it is important to briefly consider the *effects* of these policies at both the individual and societal level. Typically, economic liberalization calls for a shift from "state concentration, centralization, and equalization" to "privatization, localization, and consumer choice" (Arnove, 1997, p. 79), a move that alters the state's relationship with civil society. As the state divests itself of responsibility for social welfare, the public sphere shrinks and the individual is forced to bear an increased burden (Cvetkovich, 2012). Within this context, individual and collective notions of the "good life," whether defined as upward social mobility, job security, social/political equality, or intimacy (among others), begin to unravel (Berlant, 2011; Cvetkovich, 2012). With fewer "anchors" for the good life fantasy to attach itself to, a state of undulating *crisis* begins to take root (Berlant, 2011). And while this "crisis" that Berlant (2011) refers to is not necessarily a singular event such as El Caracazo (although it *can* be), it typically denotes a more intimate, everyday *ordinary* crisis. Berlant (2012) calls this heightened state of experiencing a threat to life, or threat to ordinariness, as the "crisis ordinariness of life." It is an impasse, a "stuckness" that is marked by precarity, anxiety, and uncertainty, all of which viewers see (to varying degrees) in *El Chavo del 8.*

Now while my intention is not to suggest that daily life in Venezuela was good and stable before the neoliberal era, the openings afforded by the Affective Turn allow me to consider economic liberalization as a circulating affective force that charged the atmosphere (Stewart, 2011) of everyday life in the country of my childhood. I posit that these policies ushered in a felt perception—a collective saturation of the senses—that an impending crisis was afoot. This was unlike the good life, or at least the "better" life that was perceived and/or hoped for during Venezuela's "exceptionalism" of the 1960s and 1970s (Ellner and Tinker Salas, 2005), a point to which I now turn.

The rise and fall of "Venezuela exceptionalism"

Unlike other Latin American countries—such as Chile, Argentina, and Bolivia—Venezuela, from the 1960s to the 1980s, did not experience a harsh military dictatorship and/or the accompanying political demobilization of social democratic policies. Nor did it live through extreme nationalist movements that led to armed conflict and economic disruption in countries such as Mexico or Cuba (Ellner and Tinker Salas, 2005). Until the collapse of oil prices and the devaluation of the currency in the mid-1980s, Venezuela had—for the most part—lived under the guise of what some historians and political theorists call the "Venezuelan exceptionalism thesis": the idea that Venezuela was actually "quite privileged"—exceptional, even—with respect to its continental brethren (Ellner and Tinker Salas, 2005; Di John, 2005). Supporters of the Venezuelan exceptionalism thesis often cite that, following the removal of military dictator Marcos Pérez Jiménez in 1958, Venezuela enjoyed political stability; the country was often touted as a "showcase democracy, an experiment widely considered successful, institutionalized, stable, and legitimate" (Lander, 2005, p. 25). The country's political system, in conjunction with its role as a major oil producer that was geographically removed from the political turmoil of the Middle East, meant that, as a whole, Venezuela did not necessarily experience the degrees of "crisis ordinariness" of its neighbors (as seen in *El Chavo del 8*).

According to numerous empirical reports, Venezuela's standard of living, or material conditions—along with life expectancy, health standards, access to housing and public services, education levels, and job employment—saw unprecedented levels of growth across the majority of the population during the 1970s (Ellner and Tinker Salas, 2005; Lander and Fierro, 1996; Lander, 2005). These favorable conditions generated high expectations of sustained improvement and "sunk deep roots in the Venezuelan mode of thinking" (Lander, 2005, p. 26), creating among many Venezuelans "the self-image of an inclusive, egalitarian, and racially democratic society" (p. 26). Optimism seemed to prevail and the "Venezuelan exceptionalism thesis" appeared to take a firm hold in the collective psyche. Popular sectors of Venezuelan society, including the urban majority of the population (consisting largely of the low and lower middle classes) arguably expected upward social mobility (Di John, 2005; Lander, 2005). With an upwardly mobile society appearing possible in the future, attachments to the good—or at least a "better"—life, seemed plausible.

This hope and optimism, however, would be short-lived. With the fluctuation of worldwide oil prices, the drastic devaluation of the Venezuelan bolivar in February of 1983, the Caracazo riots in 1989, and the subsequent structural adjustment policies enacted between 1989–1993, political, social, and, economic conditions quickly deteriorated (Ellner and Tinker Salas, 2005). Between 1984 and 1991, poverty nearly doubled, increasing from 36 percent to 68 percent (Di John, 2005; Lander, 2005), corruption became rampant, per capita income plummeted, unemployment skyrocketed, and the prospect of social cohesion and upward social mobility disintegrated. An increasingly polarized and divided society quickly emerged (Di John, 2005; Lander, 2005), with social segregation along class and racial lines becoming more salient than ever before. According to Pedrazzini and Sanchez (1994), the Venezuela of the 1980s and 1990s (which coincided with my own childhood) emerged as a "culture of urgency": crime, illegality, and an informal economy proliferated, and crisis-like conditions, which a mere decade before had seemed difficult to imagine, started to become permanent fixtures (Ellner and Tinker Salas, 2005; Lander, 2005).

Within this context, many Venezuelans found themselves in a situation of "crisis ordinariness" (Berlant, 2011), or impasse, where confidence in their world—a world they either already knew, had been promised, or imagined—began to shatter. This state of precarity, uncertainty, and disbelief, while most tangible and pernicious for the lower classes but also affecting the psyche of the middle and upper classes, made the conditions ripe for a show like *El Chavo del 8* to forge an intimate public and become an affective magnet for sizeable swaths of the Venezuelan public.

El Chavo del 8 as a site of *identification*: Scenes of crisis ordinariness

Despite *El Chavo del 8's* self-identification as a "comedy show," viewers (myself included) were presented with scenes of crisis ordinariness and its accompanying precarity. The series itself, a micro-budget production which operated under minimal financial support before achieving unprecedented success both in Mexico and abroad, showcased the lived textures of everyday life in a *vecindad*, a low-income neighborhood in an urban city center, presumably Mexico City. The show's mise-en-scène centered around the *vecindad's patio*: a decrepit, concrete courtyard inhabited by empty gas stove canisters, dilapidated bird cages void of life, window boxes planted with plastic flowers, a tenuously placed clothesline, and an abandoned wooden barrel. Within the contours of this *patio*, scenes of the banal—ordinary, everyday life—unfolded. As viewers, we would watch the "children" (portrayed by adult actors) play soccer, *la escuelita* [school], or *caminando el perrito* [walking the dog]; the owner of the tenement building walking from one apartment to another collecting rent; and women washing clothing in the community *lavadero*. Most of the characters we encountered—with the exception of the *vecindad's* teacher, Profesor Jirafales, and the owner of the building, Señor Barriga—occupied the marginal rungs of the socioeconomic strata. They included El Chavo, the show's namesake and main character, an eight-year-old orphan who would sometimes hide inside a wooden barrel as a way to "get away," and was always complaining of hunger and dreaming of ham sandwiches; Don Ramón, a single father who cobbled together odd jobs to make ends meet while nervously wondering how he would evade the rent collector; La Chilindrina, Don Ramón's daughter, who played pranks and stole toys and snacks from the other children; and Doña Florinda, a naval officer's widow and mother to Quico, who frequently grumbled about having to live among the *chusma* [rabble], among others.

Intricately entwined with these characters and depictions of crisis ordinariness, site(s) of identification emerged. As the series unfolded, Venezuelan viewers of the 1980s and 1990s could identify with the show's scenes of daily existence and characters who, despite their (mostly) "happy" dispositions, may have been also searching for ways to live a better life, or at the very least, survive daily life. The intimate public forged by the show helped viewers feel connected to others and, through humor, cultivate an affective energy that could endure the disappointment of realizing that the Venezuela they either knew, yearned for, or imagined, was out of reach. It released viewers from their "singular history" into a generality—a generality where they could become "somebody," or at least become "somebody" to others (Berlant, 2008). In a world where being "nobody" was the *modus operandi*, and where phrases such as "fue sin querer queriendo" [I didn't mean to!] or "pero no te enojes!" [Don't be mad!] were the new hello, *El Chavo del 8* provided its audience with a sense of intimacy, an "absorptive emotional world" (Berlant, 2008, p. 28) that viewers could both welcome and identify.

This is not to suggest, however, that *El Chavo*'s intimate public sustained and fulfilled individuals in the same way(s), or that the show was the sole site of identification and reciprocity among TV-watching Venezuelans of the time. Surely there were other sites of identification and recognition, including the ever-popular Venezuelan *telenovela*.[1] However, unlike the traditional, melodramatic Venezuelan *telenovela* that portrays the trials and tribulations of a romantic (predominantly white, ethnocentric, heterosexual) couple before ending in a happy union, *El Chavo* showcased scenes that were identifiable, realistic, and relatable for viewers across diverse social locations.

Middle- and upper-class viewers, for example, could potentially relate to Doña Florinda's frustration about living among the *chusma* [rabble], Señor Barriga's exasperation over Don Ramón's continuous evasion of rent, and Doña Florinda's decision to allow her son Quico to live with his rich godmother so he could get a better education[2], among numerous others. Such sights may have also been sites of affirmation: "Yikes! I don't have it as bad as Chavo!" or, "Heck! It could be worse! My entire apartment could be covered in leaks"[3]; a source of redemption: "See, [poverty] can't be that bad, look at how much fun they're having"[4] or "Look at how kind the rich landlord is! He brings gifts to his less fortunate residents!"[5]; a site of blame: "If people like Don Ramón would stop being so lazy and get a job, life wouldn't be so bad!,"[6] etc.

Conversely, for low- and lower-middle-class viewers, the show may have offered a source of consolation: "I'm not the only one who is just trying to get through the day"; solidarity, "I also know what you are going through, Chavito"; confirmation: "I, too, worry about my children going out into the neighborhood at night because of 'all the bad accidents that are happening,'"[7] etc. And for a child such as myself, the comedic, mischievous adventures of the *vecindad*'s youngsters were sights to both laugh and cheer for, particularly when—either through luck or ingenuity—the children found ways to escape the repercussions of their own antics, with the consequences often befalling the adults.

Of course, it was likely more complex and nuanced than this, but my point is that because *El Chavo del 8*'s intimate public could accommodate a range of similar *and* dissimilar *and* conflicting *and* ambivalent sites of identification, it was a capacious sustainer of people's desires, dreams, and attachments, all without forcing members into adopting the same view, reasons, or solutions. After all, incoherence, ambivalence, and/or contradictory attachments, or ways of life, are not "a sign failure, but a condition of mass belonging" (Berlant, 2008, p. 22). As Berlant (2008) observes, "to identify with someone in mass society is not necessarily to want to *be* them . . . but to be freed from being who you [are], with all of its burdensome historical determinations" (p. 29, emphasis added). By providing viewers with multiple sites/sights of identification, *El Chavo del 8* offered its intimate public a certain promise of transcendence and optimism that, as I explain next, may have actually been cruel in nature.

El Chavo del 8 as a site of (mis)recognition: (Cruel) attachment to memories of the "good life" in precarious times

According to Berlant (2011b), living requires us to be attached to something, whether that something is the fantasy of romantic love, job security, or social equality. While we might not intend those affective attachments, in order to "flourish," individuals remain attached to (good life) fantasies because they allow us to "survive being alive," despite the fact that—for so many people—those fantasies actually constitute a life that wears us out (Berlant, 2011b). For Berlant (2011), many of these attachments operate as a relation of "cruel optimism"; they are cruel because the fantasy that one often desires is actually an obstacle to one's flourishing. In the fantasies of the "good life," for example, the object associated with happiness—whether it be romantic love, upward social mobility, or job security—is actually an object that deteriorates the conditions for happiness, but its presence represents the possibility of happiness, so losing the object is deemed worse than being destroyed by it (Berlant, 2011). It is this structure of affective relationality inherent in cruel optimism that, I argue, animates not only the characters in the show, but the viewing audience as well. In the Venezuela of the 1980s and 1990s, when notions of the good life had all but vanished, viewers sustained their own fantasies through the show and through the character's attachments to the "moral—intimate—economic thing called 'the good life'" (Berlant, 2011, p. 2), even when these (cruel) attachments diminished both themselves and the characters.

The endless search for reciprocal romantic love is one such example. Although the series is replete with scenes of characters hoping for requited love (e.g., Quico and El Chavo fighting over their upstairs neighbor's affection; La

Chilindrina in love with clueless Chavo), I highlight two here. We see Doña Clotilde, for example, enamored with her neighbor Don Ramón; she is often seen scheming for ways to find him alone or embrace him, bake him cakes, invite him to the movies, or buy him a blank Valentine's Day card so that he might return the gesture. Then there is the notorious, over-the-top, romance between Doña Florinda and Profesor Jirafales—a courtship punctuated by giddy, breathless exchanges of roses and coffee and unfortunate mishaps where, at every moment of (possible) embrace, the couple is interrupted in some comedy of errors, often finding themselves either holding hands or exchanging sweet words with a different character. As a member of the viewing intimate public who was also an incredibly awkward, gangly girl with frizzy hair and a terrible fashion sense, the multiple scenes of unrequited love offered by the show were easy to identify with. As my twin sister, caretaker Berta, and I would watch the scenes of Doña Florinda and Profesor Jirafales, we would clasp our hands together and gasp with anticipation that, perhaps *this* time, the couple would actually kiss. (Our brother, on the other hand—a handsome and gregarious youngster who never seemed to lack romance during his youth—would cheer whenever Don Ramón successfully evaded Doña Clotilde.) But, in a relation of cruel optimism, when each the show's romances failed to materialize, the characters (and us, viewers) were, on some level, affectively diminished. Despite Doña Clotilde's gargantuan efforts, her love for Don Ramón remains unrequited and Profesor Jirafales never consolidates his commitment to Doña Florinda. However, rather than let go of their/our (cruel) attachments to these men and/or the notion of reciprocal, romantic love, we inadvertently placed ourselves in a situation of impasse, unable to move forward, or consider alternative objects to which we could tether our sense of continuity. Like Doña Florinda and Doña Clotilde, I continued to pine for Roberto, my decade-long childhood crush, and Berta would continue to linger by the telephone, waiting for a call from her paramour that would never come.

Another relation of cruel optimism that affectively bound *El Chavo del 8* viewers to both the show and each other was the (cruel) hope of upward social mobility and harmony across class lines. While class-marked squabbles were frequent on the show (with characters like Doña Florinda and her son Quico flaunting their perceived sense of inflated economic status and "superiority" over their "poorer" neighbors), tropes of social mobility and unity—something that was once thought possible in Venezuela—both sustained and diminished viewers. In a unique set of episodes during the sixth season,[8] viewers watched as the entire cast of *El Chavo del 8* left the confines of the *vecindad* for the seaside resort of Acapulco. Unlike the usual precarious courtyard that served as the show's ubiquitous backdrop, we watched as the characters enjoyed bountiful breakfasts together, built sandcastles on the beach, and swam in the hotel's luxurious pool. And although class tensions still peppered these episodes, the culmination of the second episode is particularly noteworthy. It ends with the entire cast joyously sitting together on the beach roasting sausages around an open bonfire with the setting sun in the background. We see the characters laughing, happily conversing, dancing, and patting each other on the back. It is an idyllic scene. As the episode comes to a close, the song "Buenas noches vecindad," written by the show's creator (and main actor), Roberto Gómez Bolaños, begins to play. The lyrics of the song are incredibly sentimental: "Goodnight my friends, goodnight my neighborhood, so long my friends, we will meet again many times, we will meet again to share our bread." While this pair of episodes is quite jarring and "unbelievable" when compared to the rest of the show, that is precisely the point. It's a fantastical construction, a relation of cruel optimism. As the credits roll, the characters—one by one—leave the bonfire, and El Chavo is seen walking off into the sunset. Like the rapidly setting sun, the fantasy of upward social mobility and social harmony quickly dissipates. In line with economic liberalization tropes of individualism—"every man for himself"—El Chavo is left alone, without his *vecindad*. Orphan or not, rich or poor, it is the individual who must fend for himself.

For viewers, the scenes in Acapulco are sentimental sights/sites of optimism and cross-class harmony. They magnetize and mobilize a fantasy of collective reciprocity, community, justice, and hope that attempt to translate the scene from the imaginary into a realm of possibility. And yet, circulating within these (and several other) episodes,[9] there is a normative subtext: social strife or social division is not caused by inherent structural inequalities, but by the failure of people to interact across class. In other words, like the characters in the show, members of different social classes can be happy and harmonious, as long as they *choose* to be. It is a matter of will and good intentions.[10] To go back to the personal, it was not until I became an adult that I began to understand the reasons behind my father's distress in the opening vignette. Growing up, my siblings and I we were allowed to watch the series every night (as long as we finished our homework), but once my brother started dressing up like El Chavo (wearing outfits with broken suspenders, floppy shoes, and ripped pants), the show was quickly banned from our household. The underlying assumption seemed to be that we could watch a show about the lower class so long as we did not become like them, a presumption that fits with Sara Ahmed's (2008) argument that notions of happiness and the good life—or in this particular case, cross-class harmony—keep their place as fantasies by their failure to be given.

While this is not to suggest that every good life fantasy is the same or that individuals in an intimate public even agree or share the same fantasy, people invest in objects that organize the relation of the fantasy to how they live, a relation that is actually cruel to their/our own flourishing. Through the characters in the show, we (as viewers) were also cruelly attached to notions of the good life (i.e., romantic love, intimacy, social harmony, etc.) in spite of their overwhelming instability and fragility in the Venezuela of the 1980s and 1990s. And yet our ongoing, affective attachment to the show (and its characters) might have tethered us to our palpably fraying world, offering a sense of "what it means to keep on living on and to look forward to being in the world" (Berlant, 2011, p. 24).

El Chavo del 8 as a site of experimentation and forging new forms of adjustment

Yet, despite my suggestion that *El Chavo del 8* was potentially cruel for viewers because it evoked fantastical notions of the good life that were no longer available (if they ever were), I contend that through experimentation, scenes of excess, and the absurd, the series presented (perhaps unintentionally) alternative forms of adjustment to the ongoing "crisis ordinariness of life" (Berlant, 2011) (Or perhaps this assertion is another manifestation of my own attachment to the show: unable or unwilling to let go, I create all sorts of justifications.)

Regardless, experimentation is an alternative form of adjustment that is interwoven in the show's production and genre. The viewer is constantly bombarded with scenes of comedic excess, creativity, and departures from the ordinary: adults playing the role of children (with hair and makeup being manipulated so that they appear ageless throughout the almost two decade broadcast); a child "coping" (with life, perhaps?) by hiding in an abandoned wine barrel; characters licking lollipops the size of dinner plates for breakfast; and bricks (or other paraphernalia) being used to bop characters over the head without ever leaving a mark. While these examples are in accordance with the series' comedic genre, humor, too, is presented as a potential mode of reparative living that can "displace" the "absurdity, drama, and afflictions" (Berlant, 2008, p. 96) of daily life.

In similar experimental fashion, through the use of the ornery and melancholic character, Don Ramón, the show inadvertently offers how embracing (rather than denying or "glossing over") "bad" affects (Cvetkovich, 2012) might actually help one come to terms with the utter disappointment, anger, or disbelief that comes with living within the realm of crisis ordinariness. Don Ramón's incessant crankiness, for example, makes him the quintessential "killjoy" (Ahmed, 2010): unwilling to let go of his suffering or adhere to societal norms regarding work, he is often seen as "killing," or dampening, certain forms of joy and the good life. If we were to analyze his character through a neoliberal perspective, we could easily classify him as a "failure": his oft-cited phrase "there is no bad work, the downside is having to work" signals a lack of will, energy, work ethic, or commitment. While Don Ramón remains poor, malnourished, and grumpy throughout the show, I contend that he is one of the only characters who manages to find traction, however slight, amidst the impasse of crisis ordinariness. Through evasion, persuasion, bargaining, manipulation, and other tactics, Don Ramón avoids paying his monthly rent throughout the entire series! Now while my main intent is not to glorify or condone Don Ramón, what interests me is his willingness to question and push back against the (normative) conventionalities of the good life (Berlant, 2008). For instance, in one particular episode, when Doña Clotilde and Doña Florinda lament over their inability to go see a movie because "Decent women should never go to a public show by themselves," an indignant Don Ramón quips, "Then why don't you become indecent?" Don Ramón's question, while posed rhetorically, can be read as an invitation to "live experimentally" instead of "living conventionally" (Berlant, 1998, 2008). In other words, he offers glimpses of what it might look like to try things out, practice new ways of surviving, and imagine new forms of being-in-relation (Berlant, 2011). This includes depathologizing negative feelings and uprooting normative attachments to societal conventions: his character insinuates (both explicitly and implicitly) how acting up, getting angry, or feeling ambivalent about one's lot in life might actually make room for conceptualizing new ideas about how to live. As one of two characters (El Chavo being the other) who periodically looks into the camera invoking the sense that he is speaking directly to viewers, Don Ramón summons the audience to question and challenge the illusory fantasy of the good life (Berlant, 2008).

And while Don Ramón is perhaps less interested in changing the world than in being defeated by it, his incitement (however small) has the potential, perhaps, to produce alternative ways of thinking about agency and change (Berlant, 2008; Cvetkovich, 2012). Such an assertion, however, requires unpacking. After all, how can one reconcile notions of agency and change in a show when, episode after episode, everything stays the same? Here, Ann Cvetkovich (2012) and Brian

Massumi's (2015) work is instructive. According to Cvetkovich (2012), impasses, or moments of not being able to move forward, are not necessarily signs of failure but may, in fact, be agentic; they require creativity, new forms of thinking and being in the world, and the creation of "stronger, more resilient communities" (p. 23). Similarly, Massumi (2015) describes how constraints, or constraining situations, actually offer constructive margins of maneuverability. He uses the act of walking to illustrate his point. When we walk, for example, we constantly deal with the constraints imposed on us by balance and gravity; we move not by avoiding these, but rather by *playing* with them. Drawing from both Cvetkovich and Massumi's ideas on the affordances offered by constraints, creativity, and stuckness, is it possible to consider that *El Chavo del 8* invited its intimate public to imagine a world where—even when things stayed the same day after day—it offered sites of potentiality? Or are Massumi and Cvetkovich's propositions simply another cluster of relational promises that are inherently cruel in nature? Like the good life that keeps its place precisely by its failure to be given (Ahmed, 2008; Berlant, 2011), the series neither arrives at a final answer nor achieves some sort of magical closure.

Conclusion: What now?

While *El Chavo del 8* is no longer broadcast on national television, it is continuously cited as one of the most popular shows on Venezuelan YouTube, enjoying perhaps an even larger audience than in the Venezuela of my youth (El Venezolano, 2016). Given its long-lasting popularity amidst the country's current state of acute (dis)/repair—where "life" itself has become literally unbearable and unrecognizable—I contend that the show *continues* to affectively bind strangers together in an intimate public that is both sustaining and cruel. Although the show implicitly and explicitly markets conventionality as both the source (and solution) to the problem of living in a world that is not on the side of anyone's survival (Berlant, 2008), at what point does the internal pressure of a frustrated promise or unattainable fantasy, unravel (Schaefer, 2013)? When is enough actually *enough*? In the messy, contradictory, collective, and affective intimate public forged by the show, might new "texts" or new openings for reimagining "life," emerge? Or, is it time to face the possibility that cruelty is inherent to any kind of affective attachment? While my ambivalence and trepidation in asking these questions might be considered a temporalized bargaining of sorts (Berlant, Tyler, and Loizidou, 2000), I can be sure of one thing: I am not ready to let go of my own (cruel) object, *El Chavo del 8*, just yet.

Notes

1 While making a causational link between the success of *El Chavo del 8* and the emergence of more "ordinary" telenovelas is outside the scope of this chapter, during the 1960s and 1970s, most of the telenovelas broadcast in Venezuela followed a *telenovela rosa* storyline, where a "poor Cinderella" falls in love with a "prince charming." Interestingly, it wasn't until the mid to late 1980s, coinciding with *El Chavo del 8's* success, that Venezuelan writers began to write more "realistic" telenovelas portraying the daily life of "common citizens." Some of these (Venezuelan-made) telenovelas included *El sol sale para todos, Natalia de 8 a 9, La señora de Cárdenas,* and *Por estas calles* (Verón and Escudero Chauvel, 1997).

2 Episode 121, *El cine* [The Movies].

3 Episode 114, *Las Goteras* [The Leaks].

4 Episode 119, *Aventuras en Acapulco I* [Adventures in Acapulco I], and Episode 120, *Aventuras en Acapulco II* [Adventures in Acapulco II].

5 Episode 124, *Festejando el día de la amistad* [Celebrating Friendship Day].

6 Episode 115, *El trabajo de Don Ramón* [Don Ramon's job].

7 Episode 194, *Intento de robo* [Attempted Burglary].

8 Episodes 159, 160, 161.

9 Some of these include: Episode 124 (*Festejando el día de la amistad*) where, in honor of "Friendship Day," Doña Florinda invites everyone from the *vecindad* over for dinner and Señor Barriga, the rich landlord, brings "friendship" gifts for everybody; Episode 126 (*El Actor Hector Bonilla*) where the famous Mexican actor stops by the *vecindad* and ends up befriending El Chavo; Episode 121 (*El Cine*), when Doña Florinda, known for warning her son, Quico, to avoid playing with the other neighborhood children because they are *chusma* [rabble], invites El Chavo and La Chilindrina to the movies.

10 As an interesting side note, these two episodes took place in Acapulco precisely because the real-life actors Roberto Gómez Bolaños (El Chavo), María Antonieta de las Nieves (La Chilindrina), and Carlos Villagrán (Quico) were embroiled in a battle over their salaries and copyright negotiation. According to several news outlets of the time, a "vacation" at a luxurious five-star hotel was the only way to keep the lesser-paid actors (i.e., Carlos Villagrán and María Antonieta de las Nieves) from leaving the production. But so much for "building bridges" or social harmony: it would be the first and only time the entire cast would appear together in any of the show's episodes (Chespirito.com).

References

Ahmed, S. (2008). "Multiculturalism and the Promise of Happiness." *New Formations*, 63.

Arnove, R. (1997). "Neoliberal Education Policies in Latin America: Arguments in Favor and Against." *Latin American Education: Comparative Perspectives*, 79–100.

Berlant, L. (1998). "Intimacy: A Special Issue." *Critical Inquiry*, *24*(2), 281–288.

Berlant, L. (2008a). *The female complaint: The unfinished business of sentimentality in American culture*. Duke University Press.

Berlant, L. (2008b). 'Thinking About Feeling Historical.' *Emotion, Space and Society*, *1*(1), 4–9.

Berlant, L. (2011). *Cruel optimism*. Durham: Duke University Press.

Berlant, L. (2011b). Public feelings salon with Lauren Berlant. Available at: http://bcrw.barnard.edu/videos/public-feelings-salon-with-lauren-berlant

Berlant, L., and Greenwald, J. (2012). "Affect in the End Times: A Conversation with Lauren Berlant." *Qui Parle: Critical Humanities and Social Sciences*, *20*(2), 71–89.

Berlant, L., and Prosser, J. (2011). "Life Writing and Intimate Publics: A Conversation with Lauren Berlant." *Biography*, *34*(1), 180–187.

Berlant, L. and Seitz, D. (2015). Interview with Lauren Berlant. "Society and Space." Available at: https://societyandspace.com/material/interviews/interview-with-lauren-berlant/

Berlant, L., Tyler, I., and Loizidou, E. (2000). "The Promise of Lauren Berlant: An Interview." *Cultural Values*, 4(3), 497–511.

Bisbal, M., and Nicodemo, P. (2006) *El consumo cultural en Venezuela. El conusmo cultural en American Latina: Constuccion teorica y lineas de investigcion*. G. Sunkel (Ed.). Convenio Andres Bello.

Cvetkovich, A. (2012). *Depression: A public feeling*. Durham: Duke University Press.

Di John, J. (2005). "Economic Liberalization, Political Instability, and State Capacity in Venezuela." *International Political Science Review*, *26*(1), 107–124.

Ellner, S., and Tinker Salas, M. (2005). "Introduction: The Venezuelan Exceptionalism Thesis Separating Myth From Reality." *Latin American Perspectives*, *32*(2), 5–19.

El Venezolano. (2016, February 29). *El Chavo del 8 es premiado como el show mas visto en youtube*. Available at: www.elvenezolano.com/2016/02/29/el-chavo-del-8-es-premiado-como-el-show-mas-visto-en-youtube/

Lander, E. (2005). "Venezuelan Social Conflict in a Global Context." *Latin American Perspectives*, *32*(2), 20–38.

Lander, E., and Fierro, L. A. (1996). "The Impact of Neoliberal Adjustment in Venezuela, 1989–1993." *Latin American Perspectives*, *23*(3), 50–73.

Massumi, B. (2015). *Politics of affect*. Chichester: John Wiley & Sons.

(N. A.) Top 10 de lo más visto en Venezuela: Venevisión acaparó el rating en el 2012. (10 enero, 2013). Available at: https://tvlatinoamericana.wordpress.com/2013/01/10/venevision-acaparo-el-rating-venezolano-en-el–2012/#more–9337

O'Sullivan-Ryan, J. (1996). *La comunicación humana: grandes temas contemporáneos de la comunicación*. Universidad Catolica Andres Bello.

Schaefer, D. (2013). "The Promise of Affect: The Politics of the Event in Ahmed's *The Promise of Happiness* and Berlant's *Cruel Optimism*." *Theory & Event*, *16*(2).

Stewart, K. (2011). "Atmospheric Attunements." *Environment and Planning D: Society and Space*, *29*(3), 445–453.

Verón, E., and Escudero Chauvel, L. (1997). *Telenovela, ficción popular y mutaciones culturales*. España: Gedisa.

Plateau III

El Chavo's *Media*-ted Encounters

7

Figures of Mexican State, Society, and Subject in Chespirito's TV

Ernesto Treviño Ronzón

Introduction: Problematizing the common

Television supplies a central reference point for the workings of a large number of contemporary societies, given that the themes and content it presents configure significant portions of the symbolic and cultural capital shared locally, regionally, and globally. This is the case for TV shows such as those produced for over thirty years by Roberto Gómez Bolaños and the media conglomerate Televisa, known around the world, appreciated for their characters, generally praised for their joyfulness and simplicity, and rarely questioned.

Evidently, neither television nor other technologies produce uniform cultural effects, nor do they merely impose content on their defenseless audiences. A number of studies over four decades have shown that, while the media has some content and specific narratives "in common," to disseminate them in popular culture, does not even remotely produce social homogeneity. Individuals and collectives mediate, resignify, and resist these messages, which are usually competing with many others.

In this sense, we are fortunately beyond studying TV in terms of a "postponed apocalypse" (Eco, 1968), or from the vantage point of structuralist approaches, although the idea of the "manipulation of the masses" still haunts (Derrida, 1989) academic debate (Sartori, 1998). We are seemingly going through an interesting time to go back to media and their contents and examine them within diverse frameworks and interrogations. This chapter examines the TV produced by Gómez Bolaños making use of some political-discursive conceptual tools (Laclau, 2005). The focus is set on the process of production of meaning and its implications for issues of inclusion, exclusion, sedimentation, and social reactivation. These tools are well known in the fields of political philosophy, sociology, and in education, but are not so familiar in communication studies, at least not in ways that are relevant to this work.

This chapter aims, specifically, to problematize some political-discursive issues stemming from the dissemination of figures of state, society, and subject in the TV produced by Roberto Gómez Bolaños, also known as Chespirito. I will anchor this debate in the field of education to provide relevant reference points.

It is worth noting that this chapter is not a positive reading of Chespirito's work, and while it does not "demolish" it, it does seek to question certain social, cultural, and political aspects of his output. I deliberately attempt to criticize some of the social and political theses produced by this kind of Mexican television, which assume a different significance as they proliferate through time and space. This is, to some degree, the task of turning these contents into rare, strange, or unfamiliar objects while at the same time problematizing those relationships between media and audience. In order to do this, I refer to concepts and contextual referents, and attempt to gain some distance—always in a limited and somewhat metaphorical way—from such a familiar and shared television experience.

This chapter proceeds as follows: first I will look at the basic aspects of the social context of the emergence and dissemination of Chespirito's TV in Mexico, as they relate to their political and cultural implications. The temporal arch is broad but not exhaustive; I refer to the 1960s and to the 1970s when his shows first emerged, then follow their trajectory to the present day, acknowledging all the limitations that such a quasi-chronological exercise implies.

I will further introduce some conceptual tools to approach the production and dissemination of media signifiers such as voices, expressions, sketches, and catchphrases. Specifically, I will tackle the notions of signification, fantasy, and interpellation, which will be examined together with political anthropology and sociology concepts such as the state and interface, to look at the differences between Mexico today and the Mexico of the 1970s. Lastly, these terms will examine the figures and symbolic offerings present in Chespirito's TV, mostly in regard to schooling and education. Given this thematic affinity, I will focus on *El Chavo* rather than the other characters created by Chespirito.

Each section should be understood as a layer to be superimposed over the rest, not as analytic transitions from theory to practice, or from context to concept, as would be the case in a research piece. Instead, the chapter is an essay that mixes and reiterates elements as a way of problematizing them. While this is mostly conceptual in nature, some references from my experience of watching the show in the 1980s and 1990s will pop up here and there. These seek to account for resistance and disagreements with the show's contents and sociopolitical functions. These disagreements are limited, as the chapter is not *strictu sensu* a discursive analysis of the series, nor does it seek to study the conditions of reception and resignification of Chespirito's TV on specific audiences.

The analysis focuses on TV, questioning its contents and gestures in an exercise of "as if": as if one were external to the phenomenon, as if one were in disagreement with it, as if the object should be different. This is a limitation, a fiction, a challenge, and a demand, as it implies producing oneself as a particular kind of author, making the cultural consumption one takes part in unfamiliar, and questioning the reference points of current popular culture.

The issue is not that popular culture is deficient or a negative mark that needs erasing, but that, inasmuch as the pop culture that Chespirito is part of is a changing collective production, its symbolic attachments and position are constantly being renegotiated. Thus questioning is inherent, not an accessory to its production, continuity, and change.

In this sense, the main thesis I will develop is that while Chespirito's TV tells sad stories with a "happy ending," it also demonstrates its constant attempts to gain a failed agency that produces systematically incomplete identities. In abstract terms, incompleteness is an ontological condition. Yet, this is not *any* incompleteness, as this show does not portray a specific society, but political-discursive productions that address and produce identification surfaces. These generate figures of state and society through which the audience negotiates its affiliations; they work as cleavages marked by violence, detachment, and poverty, in a strange spatial and temporal continuity carrying the clear effects of depolitization. These effects have been shared over many years, and need to be questioned.

Context for the emergence of Chespirito's TV

Chespirito's TV emerged in a context that is unfamiliar to current generations of Mexicans and much of its audience abroad. In terms of demographics, Mexico was undergoing a rapid transition from a rural country to one that was predominantly urban. The huge global city of over 20 million inhabitants that is Mexico City today was starting to take shape and was growing rapidly (in 1950 it had 3 million people; by 1970 that number had risen to 7 million). Narratives from that time show a country and a capital city experiencing uncertainty, expectation, growing poverty, and social marginality (Ibargüengoitia, 1990).

There was also some tension resulting from the government repression of the protests of 1968 and the following years. When Chespirito first started his work, the Partido Revolucionario Institucional (PRI) ruled the country. In none of the 31 states had there ever been any other ruling party, and this would not change until many years later.

In this context of a single-party rule, the first half of the 1970s served as preamble to the cycle of recurring economic crises that over more than 30 years would continue to affect the lives of millions of Mexicans. Partly as a result of that process of devaluation, debt, and the eventual emergence of the neoliberal state, in Mexico today, out of 120 million people, 50 million live in poverty, something that features prominently in Chespirito's TV.

When Chavo, Chapulín Colorado, or Doctor Chapatín started to become known, television was not widely available; radio still had the highest audience. There were few open TV channels, and fewer still that had national coverage. There was only one private channel, and no cable or satellite TV. In the 1970s, within this reduced spectrum, news, entertainment, governmental information, and educational programs were produced. These genres were not clearly demarcated, as is the case today when, at least in theory, political propaganda and government programs are required to identify themselves as such. There was therefore a strange mix: the certain continuity between themes and emitters of media discourses producing a particular juxtaposition. The border that allowed one to distinguish, for example, commercial TV ads from public service announcements was largely absent.

Gómez Bolaños's TV emerged in this context: a country in transition, with growing poverty rates, a single-party "democracy", media monopoly, and lacking substantial cultural counterpoints. It had great success until the mid-1990s.

Chespirito's dissemination, interpellation, and re-interpellation

Chespirito's TV produced a significantly enriched frame, as well as a series of identification surfaces. But at the same time, it was a recurring discourse, reiterative in form and content; using this imperfect yet systematic signifying organization—which produces certain signifying effects—one can order, more or less conventionally, its social and political implications.

It is worth briefly introducing some conceptual coordinates. In order to explain some of the processes of sense and meaning-making, Jacques Derrida uses different notions, such as dissemination and sedimentation, that allow him to affirm that shared reality is produced through meanings differed permanently, spatially, and temporally (Derrida, 1989). In simple terms, there are signifiers and signifieds functioning as seeds that, as they move, become disseminated and generate new signifiers and signifieds more or less close to each other.

Some signifiers produce a multiplicity of meanings impossible to map; others produce dissemination, but also sedimentation effects. Total sedimentation of meaning is impossible, because in ontological terms this would be the closure of the process of social construction. In everyday life we experience the stabilization of meanings, a certain degree of sedimentation around significant textual, oral, vocal, or graphic signifiers (Derrida, 1975). They are lived as familiarity, as ideas, themes, patterns and eventually as self-evident facts, through which we interact and produce sense; these therefore, tend to not be questioned, but rather are accepted and accommodated in place of what is already known.

More often than one might notice, some of these signifiers are the subject of controversy or dispute. They are "called into question," and this can take the form of a dispute about their meaning—what do you mean by love, democracy, equality, or freedom?—and, where applicable, this could erupt into a big political dispute. This rich set of tensions between spread and sedimentation produces social and political effects.

The TV of Chespirito is a type of discourse that for many years produced effects of dissemination and sedimentation of meanings. Programs like *El Chavo* presented a certain idea of social order and interaction; they produced a certain idea of Mexico City which was almost timeless, because almost everything was recorded in studios.

The environment created in the show as "*La vecindad*"—the "neighborhood"—the school, offices, or clinics where the gags occurred, were contextualizations, types of spatial organization little known in other parts of the country. The author of this text never came across such a place, although I lived in different parts of Mexico City for over ten years and travel there frequently. Those programs function by using simple structures, situations with dialogue, thematic discussions or exchanges that could be assumed as uncertainty, drama, intrigue, or expectation. They almost always contained aggression, jokes, disappointment, scolding, and some conciliatory outcome; possibly happy, but ultimately limited. Gestures, dialogues, and deadbolt were repetitive, and as a result, left footprints on the viewer's psyche.

Examples are expressions such as: "*Se me chispoteó*"—somehow equivalent to "it was a slip of the tongue" used by El Chavo, to give a form of explanation about some of his actions. Or "*Fue sin querer queriendo*"—"It happened but I didn't totally mean it"—which was also an explanation but settled an interesting ambiguity in the apology. "Ta ta ta ta" was an angry reaction from Profesor Jirafales prior to reprimanding or punishing any of his students, indicating that he could not articulate a word considering the facts.

The expression "*No contaban con mi astucia*"—"they didn't see my astuteness coming"—was used by Chapulín Colorado as self-praise. The character "La Chimoltrufia" in the skit "Los Caquitos" performed an impressive reiteration, using a long expression to introduce a sense of authority: "*No nos hagamos tarugos, pos ya sabes que yo como digo una cosa digo otra, pues si es que es como todo, hay cosas que ni qué, ¿tengo o no tengo razón?*": "Let's not fool ourselves, you know I may say one thing or another, it's like everything; there are things that can't even . . . Am I right or what?". Among others.

After both seeing and hearing such expressions "in action," the viewer could anticipate them with some precision. Through this process of repetition important features of convergence and differentiation of the characters happened; this was woven through their voice, clothing, concerns, relationships, and physical features.

One of the most interesting aspects of Chespirito's TV were the role models that were integrated differently and with varying degrees of effectiveness. These allowed the audience, which always included children and adults, to feel attachment, rejection, happiness, anger, shame, or sadness. Chespirito created many characters: El Chavo, El Chapulín Colorado, Quico, Ñoño, Godínez, La Popis, el Doctor Chapatin, Doña Florinda, El Profesor Jirafales, Los Caquitos. These characters functioned as caricatured representations, and in some cases openly ridiculed "real people." To some extent, they were metaphoric figures, but they were also denotative, being full of specific content, but open enough for audiences of different ages to relate to them by contemporaneity or evocation.

These characters offered specific models of how to be a child, girl, teacher, neighbor, superhero, doctor, or housewife. They also offered socialization models that guided interaction. For many who grew up watching them, gestures and words that came "out" of the screen ended up in everyday conversation, and extended to the school playground or family.

Interestingly, these characters went through very few changes throughout their television appearances. Children from the neighborhood of *El Chavo* never grew older, no one died; no central character married, no kid or grown up left the neighborhood. They were, to some extent, timeless. The recurring themes in school, the neighborhood or possibly in the adventures of El Chapulín Colorado or Los Caquitos were repeated to the point of saturation.

This last characteristic is shared by other programs, but it is important to point out that this was a specific design and requirement for television. It was designed like this and was among the first programs in Spanish spoken language to do so for many years. Its characters and situations were, to a point, their own iteration.

The dissemination of meanings derived from Chespirito's television produced a symbolic offering that can be addressed through various analytical schemes. One basic scheme involves notions of identity, identification, and interpellation. The idea of interpellation refers to the offer that a speech, a program, a text, an image makes to an "other." That other is also a subject who reproduces significations. In this case, I am referring to viewers, in a process that helps to produce identity formations.

Against some structural notions of identity formation, which tend to view identities as hard configurations derived from historically and culturally determined superstructures, in the thinking of Lacan (1983) and its political appropriations (Stavrakakis, 2010), identity is an idea used "under erasure" because it is problematic and should be questioned, as necessary, but it is insufficient for social and political analysis.

The "concepts" of interpellation and identification seem more accurate to try to address the configuration of identity formations and, where applicable, the emergence of social or political subjects. The first one, interpellation, would be a sort of "call" that interrupts the symbolic path of an unfinished subject, through content that is itself unfinished and being permanently redefined. Interpellation is the call to an identity formation from a symbolic offering. The idea of interpellation in the sense of "interruption" and "call" is related to others such as an overdetermination and displacement of meanings that Althusser (1971) recovered from Freud, and has been "updated" into the thinking of Žižek and others.

In very summary terms, this has served to account for how an interpellation presupposes the provision of a reference of meanings that can be interrupted by a call to which the subject may or may not respond. For example, for a "new" bad or magnificent performance of a play, a song, or a movie to appeal to someone, it is necessary first to know something of the original, some detail or elemental reference and depending on factors such as these, interpellation may or may not be "successful."

The results of interpellation are not entirely predictable, because they ultimately operate in the symbolic order of "shared meanings" (Žižek, 1992). In this writing interpellation is reduced to the acceptance or rejection of the call. Acceptance takes the form of a sort of incorporation of the proposed content, which in turn is constituted by a plurality of meanings around images, sounds, arguments that are received and redefined in a variety of material, cultural and psychological conditions. Studies on the television or film fruition have realized some of these rich and complex processes (Lizarazo, 2004).

The rejection of an interpellation involves the construction of a gesture of disagreement, disapproval, and even resistance, permanent or temporary. It is relevant to point out that the lack of success of an interpellation should not be understood as a simple pass-by without effects for the subject. Liminal effects emerge from affirmative or negative effects of the interpellation; these are important for forming an identity. In terms of Laclau (2005) and recovering the idea of a "constitutive outside" of Jacques Derrida and Ferdinand de Saussure, the configuration of an identity formation presupposes the construction of porous symbolic barriers, which include and exclude significant, equally constitutive elements. Rejecting a direct or indirect call is also constitutive of the subject.

Chespirito's television interpellates the public affirmatively and negatively. That largely explains its spread over several years into many territories. One of the most basic questions here is: why? The most basic answer is not because of its "quality content," but rather because it offers schemes or models that are sufficiently open and ambiguous and which are capable of appealing to different identity formations that are potentially "equivalent." That is, with similar traits and in potentially identical social conditions, at a time when television coverage was very limited.

Equivalence can be understood, following Laclau and Mouffe (2001), as the temporary process which enables individuals—or collectivities—to find a common ground or shared features, without becoming equals. This presupposes a certain suspension or partial decline of distinctive features, of differences. Equivalence does not occur in a "one-to-one" interaction. It needs a "third party" against which the different can become obvious.

In front of a television of certain kind produced in a Spanish-speaking country, which addresses in a funny or even ridiculous way the life of a public school—maybe very dysfunctional but school in the end—or a poor neighborhood—communities that are not necessarily impoverished in countries as different as Mexico, Colombia, Chile, Argentina, and Uruguay, can "converge" and become equivalent, because in all those places there are schools, towns, or neighborhoods that are not exactly equal, but share some similarities.

However, it is important to remember that the content of this type of TV, as anticipated, is sufficiently explicit and unambiguous to allow precisely that, equivalence between social contexts. But it is also important to recognize that it is not only because of what is presented as obvious or apparent content, but also for everything that is evoked, such as the underlying social, economic, or cultural arrangements. There are cultural meta-narratives clothing local narratives that the viewers then "complete" with their own contextual references. The specificity of the content is articulated with shared features and evoked traits. Again, this is neither a mere submissive transference, nor the mere acceptance that fills a void and passive receiver. It is, on the contrary, an intense act of reception and redefinition of content.

Going through the fantasy

In its most basic form, Chespirito's TV makes elementary identity proposals, pristine expressions of effective interpellatory simplicity that trigger complex processes in the viewer. On one plane they are ridiculous and anti-modelic; on another, they are evocative and also attractive without imposing any great demands. The ideological game between the characters and the audience is both rich and dynamic, and somewhat intriguing.

To speak of this relationship one could certainly return to the conventional explanation that viewers relate to the characters in Chespirito by their attractive approach, stating that the screams, blows, falls, insults, gestures, and suitable tones are more entertaining if they are mixed with naivety or tenderness. One could say, for instance, that children identify with the children in *El Chavo*, while adult viewers identify with the adults in the series, evoking as it does memories of their own childhoods, and those of their own children, grandchildren, nephews, neighbors, or students.

Without totally rejecting such readings, another will be proposed here, closer to the axis of the political; for this it is necessary to elaborate on the "subject" and to address two conceptual figures: the fantasy and the notion of subject in particular. Henceforth, some scenes from *El Chavo del Ocho*, or *Chavo*, set both in the neighborhood and at school will be used increasingly as a reference, examining some of its continuities and variations.

In a work on the notion of "event," that which encourages the emergence of radical mobility in the social, Žižek (2014) explores some theoretical familiar forms in philosophy, psychoanalysis, and cultural studies. His deliberation tackles the fantasy figure. For the author, following Lacan, fantasy among other things provides a framework that lets one experience the real as a significant whole, the world, the ridiculousness of acts of everyday reality. Through it, it is possible to support everyday triviality. In this plane and watching Chespirito's TV through this lens, it does not matter if the neighborhood does not exist, if the school is not effective or if poverty is not risible, because they are produced through a narrative framework that gives them significant coherence.

But if the fantasy is deconstructed through an event or psychoanalysis, the subject faces the possibility of de-structuring suffering, as well as the possibility of crossing the fantasy, fully identifying himself with it, functioning as a transparent background, accepting its necessity and establishing an intimate bond. Going through the fantasy or not presupposes specific types of television fruition. Its implications for the configurations of the subject are different, because while not all subjects receive television in the same way, they never do so passively; as already stated, there is always a cognitive active exercise that mobilizes a multiplicity of references (Lizarazo et al., 2013).

These conceptual coordinates are useful to think about the relationship that thousands of people have with *El Chavo* and the characters that interact in the neighborhood, or at school and with the situations in which they interact. Some use the program as a framework to make their own school environment, neighborhood or family more bearable; others use it to produce frames of interpretation of other environments by the displacement effect; still others build fantasies to consume and support the program itself and what it represents.

The content can represent an unbearable reality that viewers learned to deal with. Thousands of viewers, with more or less education, left or right, state agents or anarchists, religious or atheists, would not accept that El Chavo goes hungry, that Profesor Jirafales smoked in class, or shouted at children, mistreated students, or even paid a visit to one of the students' mothers to have a cup of coffee. They would not accept that Don Ramón constantly hit El Chavo or that he himself was attacked by Doña Florinda "with or without reason" (clearly, the idea of reason for aggression is a necessary nonsense).

Interestingly, those who reject the possibility of El Chavo's suffering re-elaborate it as informed viewers who accept the insignificance of the gag and cross the fantasy through which they enjoy it. They admit that the reality should not be like that, that it is terrible and unacceptable, but as long as it is represented in a ridiculous form, they can admit its inadequacy, elaborate it, accept and be amused by it. They manage to cross the fantasy; to enjoy how much they disapprove that a child can be struck, be starving and even be excited by the promise of an almost-always deferred piece of bread with ham. These viewers can even enjoy how much they dislike and reject the repeated gestures, the reused performative situations—Quico's sedimented cries, or Doña Florinda's facial expressions—and therefore continue to watch it again and again.

Televisa's content has been criticized because it spread social stereotypes, violence, and sexism, in an era when there were no alternative television options for a population living in poverty and economic crisis. If one wonders how this was possible, probably one of the answers is that it has acted as a mirror, or symbolic relay—to some degree "alienating"—facing terrible economic, social, and political conditions. But it was certainly not just because they helped viewers "hide" or "escape" from reality. Controversially they helped to support—even enjoy—it in the terms described above.

Here a contextual note. In a speech given in 1993 about the "world success" of a soap opera entitled "Los ricos también lloran"—The Rich Also Cry—, about the miserably unhappy life of the rich, the then-owner of Televisa, Emilio Azcarraga Milmo, said: "Mexico is a country with a very fucked modest class, which is not going to get out of it. For television it is an obligation to bring diversion to these people and to extract them of their sad reality and of their difficult future" (Villamil, 2013). This caused a stir because of the forcefulness of the owner's expression, but did not generate any surprises in terms of the company's objectives. Different actors and directors tried to clarify or re-contextualize it without much success; it eventually became a sort of negative label: Televisa produces television for people who are screwed up.

The interesting thing about Milmo's comment about this and other issues—his relationship with political power in Mexico, for example—for the purposes of this discussion, is that they can also be explained by the fact that he openly acknowledges the cynical or brazen "features" attributed to the content of his TV programmes and its place in Mexican culture. In doing so he assumes, tolerates and possibly enjoys all limitations and criticisms; to carry them on, but not through ignorance or denial. In addition to making a lot of money he can convince himself that he performs a valuable social function.

This is directly related to the emergence of identity formations which are of interest here. Both for the viewer and for the owner of Televisa, there is a process of production of meaning with ethical-political implications, where the "decision-making" regarding certain content—which is anything but innocent—is woven into dense social networks operating in the most outspoken level of "entertainment."

That is, as anticipated, Chespirito's TV works as a source of identity offerings inserted into a context of social co-production. It provides images of poverty, aggressiveness, and among these, different degrees of happiness and inter-subjective acceptance derived from the deliberate decision of a media corporation to examine a type of social and political configuration. Among other things, in this process forms of subjectivity are produced. They are articulated with state and society figures of this particular Mexican context. What are the features of this articulation and what makes them possible?

Daily production of the state in times of family TV

One of the basic arguments of this chapter is that Chespirito's television, which appeals to the family to have fun around the everyday, not only emerged in a Mexican social and political context of complex circumstances, but it also helped to produce a kind of social fabric. These relationships are references to understanding the type of television enjoyment and identification that an important part of the citizens of Mexico and other parts of the world established with it.

But there is another component of this story that is rarely explored; it is appropriate to speak about the socio-state dimension, particularly between citizens or viewers and the influence of the state on everyday life. The relationship between the media and citizens or between media and the state has been analyzed using various approaches ranging from communication, or the political, to economic and regulatory policies (Herman and McChesney, 1999; Reig, 2004). The case of media conglomerates such as Televisa and others in Latin America, their global strategy and their relationship with political power in the context of a single party has also been studied, after democratic competition and in the context of global economy (Mancinas, 2007; Mastrini and Becerra, 2006; Sinclair, 1999).

This kind of literature, somehow distant from the analysis developed here, is relevant because the persistence of television content produced by Chespirito and others must be understood as the result of several factors, including its political and structural role and acceptance by the audience. But when it comes to the people, this kind of TV has been depoliticizing, because it offers, on a global scale, family entertainment based on social problems without any sort of criticism.

A story untold in the literature, although suggested by the statements of Televisa's owner, is that while Chespirito's TV is socially and politically functional, it does not disrupt anything in social or political terms. Because of this, it was accepted by the occupying power of the state, which encouraged its permanence using various mechanisms.

In this sense, despite the advancements in the field of knowledge about TV and social dynamics, today there are still several aspects and processes which need to be addressed on a meso and micro scale. One of these is the way in which television content such as Chespirito's shares significant elements for the negotiation and redefinition of institutions, regulations, community, social or cultural structures, and how they help, in turn, to support the interaction between citizens and the state. Or in simpler terms: how they help to produce a certain type of statehood through daily interaction.

To address this, a brief conceptual detour is necessary. In this regard, the recent Mexican context maintains lines of continuity with the context in which Chespirito's TV emerged. Since that time, the country has experienced what has been called a crisis of the state, an expression grouping processes of different scope, for example, the inability of institutions and their agents to fulfill their responsibilities, such as establishing a functional rule of law, reducing levels of violence, poverty, marginalization, and the provision of public services. Several of these crisis markers present during the rise of Chespirito still remain, and in the last decade of the twentieth century, corruption and impunity were added (Zepeda, 2004; Le Clercq and Sanchez, 2015).

Recent analyses of the Mexican context have shown repeatedly that these problems do not show just corrupt or ineffective rulers, or badly designed policies, but also the type of citizens and groups inhabiting the territory that help in the co-production of the social and state formation. That is, the narrative of the state crisis has been completed and overlaid with a newer one concerning "absent" citizens, which have historically sought parallel, informal, and even illegal solutions to solve their problems, creating effects that disable the creation of collective benefits and modification of the state.

This would include state agents themselves, but also parents, youth, entrepreneurs and traders, doctors, teachers, lawyers, people in communities, neighborhoods and big cities, to mention just a few, who would participate in an intense process of retrofitting rules, structures, and state boundaries. It would be a permanent inner process, not an accessory to the crisis of the formal state, which would have been semi-hidden behind the fiction of the full presence of the state throughout the territory (Abrams, Gupta, and Mitchel, 2006). Thus, the initial narrative is corrected: in the case of Mexico the state functions deficiently but state and non-state actors themselves systematically collaborate in this process (Selee and Santin, 2006; Merino, 2013).

In the case of education, this crisis presents itself in various ways (e.g. in basic education, in the deterioration of school facilities, the decrease in public spending on the training of teachers or opening new schools). Also, in the expansion and influence of the National Union of Education Workers (SNTE), the largest union in Latin America, who for many years controlled the hiring, promotion, retirement, and even the political affiliation of the millions of teachers that are its members (Ornelas, 2012). This crisis in education is explained not only by the failure of formal state institutions, but also by the absence of citizens who, in their diversity and incompleteness could claim, be pressed or forced to help concretize the right to education expressed in the Mexican constitution.

Of course, one cannot ignore that in Mexico from the 1970s up to the present day there have been several arenas where citizens have tried this. There have been places where they found and exchanged different "social worlds," where interactions have revolved around the disputed views that would eventually guide actions, where the presence of the state has been mostly figurative, activated and deactivated in everyday social interfaces by the people (Long, 2001).

Student movements such as Zapatismo with its project of political, economic, and educative autonomy are another example. On a different scale, there are also examples of how citizens interact with environmental policies, actions and institutions of health, safety and education (Isunza, 2004; Hevia, 2009). A public school is one of the richest places in Mexico to see how the state is produced in everyday activities. This is so, because they are grounds for dispute and the negotiation of imperfect rules where laws, policies and programs, management and management styles, aspirations of parents, the specter of curricular standardization, various representations of school participation and democracy collide, are blocked and traded in an overdetermined infinite play (Treviño, 2014).

But the more one delves into the history of the state crisis, the harder it is not to include in the analysis the impact of mass media. While this is a discussion that involves dealing with a dense network of relations impossible to detail here, it is possible to pull some strings. Chespirito's TV gives us something relevant because its contents must not be analyzed as mere "portraits" of society, but as an inner member for the co-production of a type of Mexican society and state.

El Chavo and other characters throughout the years have a curious and controversial role; they have served to provide those fantastic matrixes discussed in the first section of this chapter. In the same process they have served to produce a type of symbolic and political interface, through which forms of relationship between viewers, social problems, and the actions and inactions of the state are generated.

Specifically, the type of television that Chespirito produced helped to create a basically apolitical treatment of the social. It is not the kind of apolitical attributable to alienation or the denial of problems. It's the kind of apolitical that is rather problematic as it passes the reality of poverty, hunger, aggression, insecurity, ignorance, and lack of education, as funny, sad problems that are yet livable and enjoyable. It is the type of content that, while showing the reality, does not pose a cognitive, social, political, or cultural conflict beyond the gag, because in the same process any possibility of conflict is disabled.

There is an effect of systematic discursive decline which is also a political decline. A look at some expressions would be useful. Looking at the following phrases, habitual viewers would remember the gestures, tones, and body movements that accompanied them (see the very rough translation next).

- From El Chavo: *Es que no me tienen paciencia*—You have no patience for me.
- From El Chapulín Colorado: *Se aprovechan de mi nobleza*—They take advantage of my nobility.
- From El Chavo: *Fue sin querer queriendo*—It happened but I didn't totally mean it.
- From many characters and skits: *Chanfle!*—No translation (surprise, wonder).
- From El Chapulín Colorado: *Bueno, la idea es esa*—Well, that's the idea (used when a saying could not be completed).

Observed from a discursive-political approach, each of these expressions has a declining hue. The first and second are openly pleading for calm, forgiveness, and understanding for what could be a failure or a greater expectation. The third, fourth, and fifth show some disruption: the first part of the formulation indicates that something happened and that something else can happen, there is a kind of uncertainty. As Derrida (1982) indicates, in the path of meaning there is time and space. In this case it serves the explanation: what almost automatically follows that small opening, almost imperceptible to the audience, is the decline, a sort of cancelation of the disruptive effect. Nothing disruptive happens later, because what happens is comforting laughter.

Apply that to "examples" of everyday life, moments of citizen–state interface. After many twists and turns an unemployed citizen finally has enough money to pay his energy bill; after running, jumping, pushing down the street, when he arrives at the window he finds the computer is not working, or the bureaucrat will not accept cash, or for some reason he cannot pay. After days without money, furious with both himself and the world, he must come back tomorrow or Monday assuming they have not cut off the power in the meantime.

Another sick citizen waits at the public clinic, until he is at the point of desperation. The nurse sends the patient home with the promise that tomorrow the doctor will see him and that there will be medicine or, in the worst-case scenario, leaves him uncertain as to whether he will be taken care of. After a mini-crisis, the subject accepts and assumes his own conflict because ultimately, he is still sick and he has to come back.

Any possibility that these citizens can escalate their frustration is, for now, in the course of this story, canceled. Or evoking Rancière (1999), there is no possibility that these subjects—for decision or effect—question their place in the scheme of things, there is no way that the conflict is installed: a systematic questioning of the state of things that scales to a permanent conflict where power relationships and attributed identities are rejected and renegotiated. Instead, what we see is the victory of administrators.

These are forms of daily production of statehood: they are not the only ones, and they are not generally of interest to political scientists or statesmen. They are of interest to those who, in different disciplines, observe the common discursive creation and negotiation of rules, limits, margins, conflicts. Also self-containments, individual and collective implosions which ultimately produce forms of state and society. Chespirito's television produces such discursive forms; it helped to create a cultural matrix that is also depoliticizing.

The school of *El Chavo*: Social continuity and the obturation of the state

This section introduces some derivations of the ideas previously examined in conceptual and contextual terms. It is not an analytical argument but rather a focused reflection. In Chespirito's television, we see what could be called a permanent negotiation of central ideas of coexistence and interaction in contemporary societies, such as authority, legality, normality, solidarity, and companionship.

The most obvious case, because of its design as social interaction, is *El Chavo*. In this show conventional ideas, such as the nuclear family, are displaced and "do not exist"; those who watch it may notice, even *a posteriori*, and possibly through the grid of a gendered discourse, that in the neighborhood there are no "traditional" nuclear families with a dad and a mom. You can also note that except for the incidental appearance of any police or the mailman, state representatives are not listed as such; the most obvious is Profesor Jirafales, both inside and outside school.

In the *vecindad* one can witness the configuration of a specific form of social regulation, which means that local agents set their rules, negotiate their time, set their limits, and price the nature of their actions and the corresponding sanctions. Probably for this reason, throughout the entire duration of the program, the orphan child from the neighborhood lives in a barrel. No one seems to bring him to an orphanage, and it does not seem to be a problem that he lives in potential risk of ill health, without food or worse. That's probably why violence between neighbors has never escalated to a lawsuit.

Now, viewers of *El Chavo* have a "special" relationship with the representation of the school, a place that we both miss and detest. It is possibly the best surface to read the inscription and negotiation of boundaries between state, society, and individuals as noted in previous paragraphs. In *El Chavo's* school, apparently no one learns. It is an anarchic cosmos, where children do not do their homework, and even if they do, they are usually wrong, and so do not learn on their own. In this school, peer violence prevails, interwoven with ignorance, cheating in exams, lies, and a general unwillingness to study.

It is more than evident and it "would be expected" that children from the *vecindad* would be undisciplined at school. And that would be the reason for Profesor Jirafales to easily move from his initial Socratic reflexivity, to become an entity that shouts, throws fits, and smokes constantly. Subjected to this chaos, where it would be impossible to work, at some point, pedagogy does not exist in the eyes of the TV camera; it is suspended as a school's discursive element.

Any act of mediation or educational guide is secondary and "understandably" attempts at functional control come into play: repetition, shouting, scolding, pleas. These in turn are not successful, but become a routine which frequently ends with the teacher's trauma as he abandons himself to the chaos of the classroom.

Again, there are several ways to read this little narrative so characteristic of *El Chavo* and replicated in many other television programs in Mexico. One is that this representation is a true and accurate representation of thousands of public schools in Mexico and possibly in other parts of the world. As such, it is plausible to look at it and laugh without further thought: this is reality and nothing can be done about it.

Another way to see it is through a more elaborate lens. From a such a position, one could say that *El Chavo* presents a devastating critique of the educational system, all its problems, all its faults, its real inability to achieve the "promises" of modern education: guidance, the overcoming of ignorance, the production of enlightened subjects, made concrete through an undeniably apostolic educational process.

From another perspective one might even say that it is possible to see a fine, albeit cartoonish, representation of the mediocre Mexican state schools of the period, which precisely because of that became a state in crisis. In this sense, it would be a faithful representative of the Mexican school age, with a state agent in an equally suitable classroom: the anti-pedagogical Profesor Jirafales who, in his inability and frustration can do nothing but fail. Those who enter the school, students and parents, dislocated subjects with the need to know, leave no less dislocated and without learning anything. They are failed identities in the sense of open and incomplete, and this would be wrong. Both readings are plausible, but have a variant. In it, this school is a caricature of "what is," something that needs to be laughed at. At that failed school, for these low-class children living in an ugly neighborhood, with no great economic and cultural resources, who couldn't even think of complaining, reality is symmetrical. They cannot get out of there; their parents do not expect much of their education, so they have the perfect school.

The children in "*La vecindad de El Chavo*" live in dysfunctional, disjointed, mediocre, and aggressive families, which produces no agency; in these families, gossip prevails, adults do not work, and they hide from the landlord to avoid paying the rent. All this is the subject of comedy. Not having family or religious or political or artistic agency, why should school be different? Why should any of this scenario change? Justifiably, because of this, in this school there is no room for an imaginary of this nature, there is no necessity for a salvation narrative; even proposing one would be nonsense, because it would prevent millions from crossing through and escaping the fantasy.

What counts here are the small individual satisfactions, expressed in some micro recognition, fleeting, sporadic, not resulting from an individual or collective effort, but from the fault in the other, the one who is scolded, beaten, punished, even exposed. In the social organization presented in *El Chavo* there is a clear continuity between the social structure of the neighborhood—the *vecindad*—and the social and cognitive structure of the classroom. The school is for the neighborhood and vice versa in a kind of ontological alignment.

Seen from this angle, what might seem like a devastating critique embodied in a show, becomes over the years a form of sedimentation. This discursive model and its content is spread and fixes with some forcefulness a certain idea of school, for a certain type of girl and boy, in a certain kind of social history, in a timeless context.

Of course, it is possible to ask if this story could not be different, if that program could not have had a better school, a better teacher, better classes, more intelligent children or even another form of social structure, and still make one laugh. One might wonder whether in that school Chespirito could not have thought of producing "successful" identities; different, a little rounder, less "failed" assemblages, producing "happy" children, "satisfied" parents and teachers. Why not have adults learning with their children, not beating them or insulting anyone? Why not present the strange combination where the poor kid in the neighborhood is smart in school? Why not present the one that despite poverty or violence returns with stars, congratulations, and knowledge?

Interestingly this narrative of salvation was not present; we never saw the emergence of the "novel of formation," where subjects change over time in the eyes of the viewer becoming a "smart" boy or girl who can overcome obstacles. We witnessed only small achievements, pyrrhic satisfaction with the successes in error, or sometimes by knowledge, but always disqualified. Like when Ñoño gave a correct answer to the teacher and he was immediately dismissed by his companions, as if it were the only thing they could do.

In the transition of the above-mentioned explanatory possibilities the spectator faces the challenge of doing something with this school. As already said, viewers are not passive; we are active in front of the TV: we can reject it, laugh with its literal content or pass through fantasy, becoming a kind of ethical–political subject, accept it and enjoy it.

Options like ignoring it were difficult at the time, less plausible was fighting it, they were never options; it is not possible to know if they are now because we do not know of any public voice that has said that such a content presented for so many years should or must be rejected.

There is still a great deal of affection for Chespirito and actually his work was never seen as something that should be questioned. It is not until recent years that some voices began to consider these issues and it is difficult to know if any of them reached his ears.

This school, this form of social arrangement is a stable production, a regime of meaning that as a result of reïteration produces a specific form of statehood, which despite appearances, nevertheless cannot be "closed" in ontological terms.

This proposal approaches the viewer as a discursive matrix through which one can create relationships, deal or not with problems, relate to the state or build it in everyday ways that are largely depoliticizing. The Mexican State offers schools that, if they are not like that of *El Chavo*, seem a bit like it, and in that sameness can be endured.

Final comments: Does Televisa produce apolitical subjects?

Created by a transnational company with wide penetration in the domestic and global market, Chespirito's television served different purposes, including those of political order in scales micro, meso, and macro social. At a macro level it was functional, willingly or not, to the consolidation of a one-party state for more than 30 years. It was the star of primetime for a media conglomerate that openly avoided questioning the exercise of political power and its most obvious consequences at the time, though it wasn't, strictly speaking, state television.

In terms of the representation that the company makes of itself and its contents in the voice of its owner, they functioned as "soldiers for the PRI and the president" (Villamil, 2013). One must consider this sentence and ask if it really meant anything, or whether this was an exaggerated self-attribution. Perhaps Chespirito's conduct was "politically correct" at a time when this was not needed. While there may be several ways in which a media company can relate to the political, economic, or religious power, in this case, for a long time there was only one: structural functionality. And there is no evidence at hand to say otherwise, although, as discussed later in this section, some recent variations have been observed.

At the meso-social level, Chespirito's television strategy served, among other things, to disseminate an idea of organization and media interaction, and ultimately social simplicity based on stereotypes. At the micro-level, the social content of Chespirito's programs rarely addressed situations that involved a substantial ethical dilemma, a "moral learning," a collective dispute, a bond of solidarity or an example of social democracy.

The closest to this was reflected in the comic figure of El Chapulín Colorado, within which it is possible to identify the prevalence of small satisfactions. And quite possibly, this oft-repeated content, expressed through identity models sufficiently full and properly empty, supported the belief that, given the impossibility of obtaining "states of lasting happiness," or satisfaction in a framework of rights and democratic opening it was convenient to "laugh."

Laughter, satire, irony, like art in its different manifestations—at home, on the street, at school, in the parliamentary floor, in the press, on TV—have an interesting role in some contexts and it is essential to destabilize the prevailing political or cultural order. However, in Chespirito one cannot find this kind of laughter; it therefore should be criticized because it is closer to a deactivating gesture and its scope is limited: for many reasons it makes you laugh, but not because of relevant political reasons. It is the laughter "that helps" the viewer to bear everyday life, not to question it.

This can certainly be very important for millions depending on the context—bearing everyday life with the help of TV—but it is not admissible to assign a universal value to this function of TV content.

There was a subtle change in the political function of Televisa when the PRI left the presidency of Mexico in 2000 and the single-party regime changed. Chespirito interacted frequently with leaders of the center-right National Action Party (PAN) particularly President Felipe Calderón. True to his own logic to stay close to political power, Gómez Bolaños helped the regime to develop and spread a conservative idea of family values during the six years from 2006 to 2012.

At the same time there was a brief differentiation process in Televisa. During the latter part of the Calderón administration and near the death of Gómez Bolaños, the company began to criticize the PAN and the Mexican government. It directed criticism at specific sectors such as public education.

Emulating the documentary "*Waiting for Superman*," released in the United States a few years earlier, in Mexico the documentary "*De panzazo*" (Barely Passing) was recorded and became widely known, narrated by one of the most popular anchors in the country. It was produced by a private organization called Mexicanos Primero, which was funded by the owners of major multinational companies.

The documentary and the treatment of political issues by Televisa started an interesting battle. It openly criticized the state of public education in Mexico using questionable information and data, based on simplistic arguments, openly discrediting teachers, parents, and the government, but without assuming any responsibility for the role that the media had in spreading socially and culturally controversial content.

This same alliance pushed an agenda that would later become a proposal to reform the public education system of Mexico reflecting the program of the new PRI president, Enrique Peña Nieto. He has long been a political figure associated with Televisa shows.

Today, Televisa faces competition in the media market; in fact Mexican families can watch dozens of open and closed channels, which has undermined the impact of the conglomerate. Partly due to this process of opening in the media market, which occurred during the PAN administrations, Televisa chose to support the PRI. Today Chespirito's show is not the most watched. The production of *El Chavo Animado*, accompanied by content in social networking platforms and with a range of products associated with it have tried to maintain its popularity, but the map of sociopolitical interaction of these contents has inevitably changed.

For many people in Mexico (and other countries) who grew up watching programs like these, in a context of economic crisis, poverty, and a single-party state in crisis, the symbolic link to Chespirito and his characters remains strong. Considering this and its consequences, it is mandatory to explore the scope of this cultural matrix, of these models of identification and production of state and along with that, the potential for undermining them.

TV has no power to impose its will on the identity configuration of political subjects, but it is possible to argue, based on a look at its content model, that depoliticization processes operate there. So putting aside romantic evocations and candid laughter, it is necessary to examine the depoliticizing effect that this type of television had on several generations of Mexicans.

It's not about ascribing powers and disproportionate consequences, but it is essential to ask questions along these lines, and place them in a broader network of relationships that, at least in Mexico, allows the field of analysis and interpretation to expand. While these contents produce that widely spread social micro effect, Televisa has never stopped playing a role in the great political arena.

This chapter is a contribution in this line of reflection that in Mexico and other parts of the world tries to introduce some variation in the field of political debate. It tries to redraw the idea of media coming from the left and right through a conceptual and analytical reprocessing (Strathausen, 2009). It is important to produce scenarios of political action and alternative social participation, in an age where technological density increases and is permanently present in complex processes such as education, inside and outside schools. It is undeniably a theoretical, technical, political, and ethical task.

References

Abrams, P., Gupta, A., & Mitchel, T. (2006). *Antropología del Estado*. México: Fondo de Cultura Económica.

Althusser, L. (1971). "Ideology and Ideological State Apparatuses." In L. Althusser (Ed.), *Lenin and philosophy and other essays* (21–176). New York: Monthly Review Press.

Derrida, J. (1975). *La diseminación*. España: Fundamentos.

Derrida, J. (1982). "La Différance" (1–27). In *Margins of philosophy.* Sussex, England: The Harvester Press Limited.

Derrida, J. (1989). *La escritura y la diferencia*. Barcelona: Anthropos.

Eco. U. (1968). *Apocalípticos e integrados*. Spain: Lumen.

Herman, E.S. and McChesney, R. (1999). *Los medios globales. Los nuevos misioneros del capitalismo corporativo*. Madrid: Cátedra.

Hevia de la Jara, F. (2009). Relaciones sociedad-Estado: análisis interactivo para una antropología del Estado. *Espiral,* May–August, 43–70.

Ibargüengoitia, J. (1990). *Instrucciones para vivir en México*. México: Booket.

Isunza Vera, E. (2004). *El reto de la confluencia. Las interfaces socio-estatales en el contexto de la transición política mexicana (dos casos para la reflexión)*. Xalapa, Universidad Veracruzana (Cuadernos de la Sociedad Civil, núm. 8).

Lacan, J. (1983). *El yo de la teoría de Freud y en la técnica psicoanalítica*. Buenos Aires: Paidós.

Laclau, E. (2005). *La razón populista*. Buenos Aires: Fondo de Cultura Económica.

Laclau, E. & Mouffe, Ch. (2001). *Hegemony and socialist strategy: Towards a radical democratic politics*. London: Verso; 2nd edn.

Le Clercq, J. A. & Sánchez, G. (coordinadores) (2015). *Índice Global de Impunidad.* México: Fundación Universidad de las Américas Puebla.

Lizarazo, D. (2004). *La fruición fílmica: Estética y semiótica de la interpretación cinematográfica*. México DF: UAM-X.

Lizarazo D., Andión, M., Hernández, G., Hernández, D., Treviño, E., and Millán, M. (2013). *Símbolos digitales. Representaciones de las TIC en la comunidad escolar.* México: Siglo XXI Editores y UAM.

Long, N. (2001). *Development sociology actor perspectives.* London: Routledge.

Mancinas, R. (2007). El desarrollo de grupos de comunicación en México: el caso de Grupo Televisa. *Razón y Palabra, 12* (59), 13.

Mastrini, G. & Becerra, M. (Eds) (2006). *Periodistas y magnates. Estructura y concentración de las industrias culturales en América Latina*. Buenos Aires: Prometeo.

Merino, M. (2013). La captura de los puestos públicos. *Revista Mexicana de Ciencias Políticas y Sociales*. Nueva Época, LVIII (219), 135–156.

Ornelas, C. (2012). *Educación, colonización y rebeldía: La herencia del pacto Calderón-Gordillo*. México: Siglo XXI.

Ortiz Leroux, S. (2010). La crisis del Estado mexicano: una lectura desde el republicanismo de Maquiavelo. *Argumentos, 23(64),* 37–60.

Rancière, J. (1999). *Disagreement. Politics and philosophy*. Minneapolis: University of Minnesota Press.

Reig, R. (2004). *Dioses y Diablos Mediáticos. Como manipula el poder a través de los medios de comunicación*. Barcelona: Urano.

Sartori, G. (1998). *Homo videns. La sociedad teledirigida*. Spain: Taurus.

Selee, A. D. and Santín del Río, L. (2006). *Democracy and citizenship citizen participation and public deliberation in Mexican local governments*. Washington, DC: Woodrow Wilson International Center for Scholars.

Sinclair, J. (1999). *Latin American televisión. A global view*. Oxford, UK: Oxford University Press.

Stavrakakis, Y. (2010). *La izquierda lacaniana*. Buenos Aires, Argentina: FCE.

Strathausen, C. (Ed.) (2009). *A Leftist ontology. Beyond relativism and identity politics*. Minneapolis, MN, USA: University of Minnesota Press.

Treviño, R. E. (2014). "El programa escuela segura en escuelas de Veracruz: condiciones de recepción, matrices de representación y sus implicaciones para la participación social." In A. Zavaleta, E. Treviño. & M. Jiménez, *Elementos conceptuales y metodológicos para el estudio de la violencia en las escuelas* (85–105). México: UV, CONACyT.

Villamil, G. (2013). Televisión para jodidos, Semanario *Proceso*. Retrieved on April 29, 2016, from: www.proceso.com.mx/336733/television-para-jodidos.

Zepeda, G. (2004). *Crimen sin castigo: Procuración de justicia penal y ministerio público en México*. México: Fondo de Cultura Económica y Centro de Investigación para el Desarrollo.

Žižek, S. (1992). *El sublime objeto de la ideología*. México: Siglo XXI.

Žižek, S. (2014). *Acontecimiento*. Madrid: Sexto Piso.

8

Media Education "*Sin Querer Queriendo*"

Dulce María Cabrera and José Carbajal Romero

El Chavo del 8, which aired for the first time on June 20, 1971, was a weekly TV series broadcast every Wednesday evening at 8 p.m. on Mexico's Channel 8. From its inception, the show was considered a media event of great importance and transcendence, particularly around figures and meanings related to the urban-popular. The renowned program portrayed many of the emblematic forms of poverty that, even during the 1970s, were still observable in the *vecindades*,[1] or housing projects, of Mexico City.

Scarce scenic resources and dramaturgical talents combined to create a media product of extensive educational impact that, while reiterating the clichés of social and cultural practices of low-income settlements, showed the dynamics, social relations, collective memory, linguistic practices, and clothing styles of an urban context, and the extreme poverty of the social sectors that had sprung up near the capital city due to increased migration from the countryside. All of these, *sin querer queriendo* [without meaning to] profiled some of the symbolic identification processes that were made possible to the show's viewers, including, but not limited to: the adoption of particular characters, the incorporation of phrases like *¡Me lleva el chanfle!* [Damn!], or *¡Se me chispoteó!* [It just slipped out!] into the daily vernacular, and the adoption of particular fashion styles, etc.[2]

To speak of *El Chavo del 8* involves an exercise of reflection on the relationship between the urban-popular, cultural reference "frames," and mass media. It involves analyzing the educational from a contingent logic, one that does *not* depend on the interpellations offered by the content of the series, and does *not* respond solely to the will of its subjects. This series, like any cultural product, has educational implications beyond the intentions of its producers or the modes of identification suggested by the program. They are (educational) components that emerge in an intermediate zone that is neither determined nor determining.

Beyond any particular intention, the media education developed through the series *El Chavo del 8* occurs "without meaning to," *through* the media, *in* the media, and *against* the media.

While we recognize that education can take place via several technological platforms, we believe that television transmits certain cultural reference frames that denote multiple forms of identification. In this vein we argue that media education developed through *El Chavo del 8* creates an aesthetic experience *through* the *vecindad*, or housing project. However, to argue that education takes place *in* the media, would (at first glance) imply the simplistic idea that viewers are replenished solely by the codes provided by television. But this is not the case. Between transmission and identification, links that tie the construction of hybrid figures (whose ingredients contain elements of popular culture and the subjective experiences that become conjoined in "unexpected" ways) occur. In this regard, we believe that what is educational *through* the media is a process that allows individuals to construct the figure of El Chavo as well as resignify their own subjectivity.

However, we recognize that the links between the creative, formative process and the reference markers deployed by the mass media are not necessarily directed toward their typical audience expectations. While it is important to recognize that transmitted content can have ideological effects, we believe that educational processes are never unidirectional. That is, media education does not just happen *from* the medium *to* the subject; there is always the possibility that processes of resignification go *against* the medium.[3] What is educational about going *against* the media is that it alludes to a political force that involves particular dimensions of television and generates unexpected results. According to dominant cultural and sociopolitical logics, the remote transmission of cultural, ideological, and political content on a mass level is insufficient to determine the production of subjectivities.

In *El Chavo del 8* we see that there are disruptive, unforeseen processes that can alter the symbolic plot offered by television. Here, we would like to point out that we are interested in showing how the identifications the public makes with El Chavo are not merely the result of direct consumption of the show's content, but also involve subjective configurations that cannot be foreseen, such as the the adoption of the "marginalized" as a way of life (something which could never have been predicted forty years ago[4]). In the last part of this chapter we offer our conclusions, insisting that media education happens "there," where you least expect it, *sin querer queriendo* [without meaning to].

Media education

Analyses of the interactions between subjects and technologies—particularly those related to media—have been conducted from different angles (Eco, 1984, McLuhan and Powers, 1995; Sartori, 1998):

> For several decades, technological processes associated with the processing and sharing of information have opened the debate around the uniqueness of the educational, giving way to more expansive processes that decentralize education from the school space and take it to other social agencies and organizations that are distant in both origin and function from the processes of teaching and learning that occur in schools (Carbajal, 2003, p. 56).

Our work inscribes itself in this debate and argues that the educative involves a process of formation; it is an original production that escapes the will of the subject and articulates references that come from different orders. Their configuration involves a political exercise in which significant forces and elements compete and are combined with other subjective processes such as identification and recognition. We would argue that "the educative constitutes and institutes the space of the formation of subjects; as such, it is a political space where processes of identification may occur" (Carbajal, 2003, p. 57).

In this process there are several competing "frames" of reference that affect identification. These "produce a historically contingent ontology, such that our very capacity to discern and name the 'being' of the subject depends on norms that facilitate that recognition" (Butler, 2010, p. 17). For example, in *El Chavo del 8*, the *vecindad*, or neighborhood, becomes a stage for comedy and at the same time, creates a reference frame that contains elements related to poverty, misery, urban life, and childhood. The origin of this cocktail of significations, however, does not appear explicitly in the program; television promotes its reproduction and turns it into a possible identification marker: the droll anecdotes that we see between El Chavo and his friends, for example, create a frame of reference for the rest of the inhabitants of the *vecindad*, representing all of them as poor, marginalized, or miserable.

Intermezzo

Throughout this chapter, we argue that television provides reference frames that are combined with *other* elements of the sociopolitical environment that do not translate from the medium to the subject in automatic, intentional, or unidirectional ways. This is due to the fact there exists a space between the transmission of content (the broadcast) and a subject's identification. Our analysis privileges the *intermezzo*, where the unintentional connections that give meaning to the educational are "always between two points, but the in-between has taken on all the consistency and enjoys both an autonomy and a direction of its own" (Deleuze and Guattari, 2006, pp. 384–385). That is, the educative is the *intermezzo*; herein lies the importance of media literacy: it happens *in* the media, *through* the media, and *against* the media.

Education *through* the media

In our approach to media education, we are greatly interested in the educational modulations that can arise from a cultural product like *El Chavo del 8*. This is based on de Certeau's (2000) argument:

> The presence and circulation of a representation (taught by preachers, educators, and popularizers as the key to socioeconomic advancement) tells us nothing about what it is for its users. We must first analyze its manipulation by users who are not its makers. Only then can we gauge the difference or similarity between the production of the image and the secondary production hidden in the process of its utilization (p. 43).

Our research is situated within this difference. The difference that de Certeau discusses—between the production of the first image on television and the secondary production that takes place between subjects—can be explained by the enormous gap that exists between the cultural content intentionally launched by television and the appropriation that "viewers" make of these (as part of the processes of identification). As such, it follows that the images produced by subjects are not the result of the content's "direct consumption" (although we recognize that this can affect their creation) but may be the result of contingent social, cultural, and political factors. This means that at least two images are constructed, one that corresponds to El Chavo as a character in the show and the other in the viewer, a *chavo* (or "lad") as an indeterminate signifier.

De Certeau (2000) notes the lack of similarity between the two images and indicates that a process of manufacture and manipulation occurs on behalf of the subject. In this regard, we contend that the image transmitted by *El Chavo del 8* maintains a distance with the image constructed by its viewers because a gap exists between the show's content and the identification process of the viewers. In other words, the space that exists between the (televised) image and the subjects supports the combination of multiple heterogeneous elements. It is in this intermediary space where the educative occurs; it is in this space, coalescing around the figure of El Chavo, where ethical-political-aesthetic elements come together.

However, it is important to emphasize that the connections between the character and the viewer do not simply come together in the intermezzo. For, as de Certeau (2000) argues, it would be a mistake to assume that a subject's only task is to identify him/herself with what is presented. The appropriation of cultural elements that become re-iterated[5] is never absolute; subjects do not simply adopt or reproduce all that *El Chavo del 8* offers in its totality.

We believe that the *intermezzo*, or the space between the production of the television series and the subjects' identification processes, is thick and viscous; it is not simply a matter of crossing the two to produce some sort of isomorphism. There are no unique and harmonious combinations, but arbitrary and unintended hybridizations; rhizomatic connections where the most unexpected identifications can occur. We can see this in the way that over three generations of *El Chavo del 8* viewers have adopted some element of the show as a cultural reference, but leave out many other elements capable of generating identifications.

Who is El Chavo del 8?

Undoubtedly the most complex and enigmatic character in the *vecindad,* El Chavo del 8 managed to arouse sympathy among his viewers. While part of the character's reception was the result of the program's structure and guidelines in Gómez Bolaños's scripts, an important element was the character's ability to summon different meanings and create identifications across various social groups and nationalities.

El Chavo, a poor eight-year-old orphan who wears the same striped shirt and tattered pants with patches throughout the entire series, is frequently seen with a charming, disarming smile. His lack of material goods and impoverished state are often accentuated to highlight symbolic elements such as innocence, goodness, and the joy of childhood. The figure of El Chavo highlights stoicism: he is noble and overcomes adversity without rancor. This is evident in the daily adventures with La Chilindrina, a bold, restless child who is often seen instigating mischief in the *vecindad.* In numerous episodes, La Chilindrina seeks El Chavo's affection, but if the situation ever calls for her to throw her accomplice under the bus in order to save herself, she never hesitates, and can be seen sympathizing with El Chavo by saying "*ay chavito, chavito!*" (Little lad, little lad).

El Chavo is an urban hero. While not superhuman, he exhibits the human powers of empathy, compassion, courage, and fights for those who are suffering even though he is often the one who faces the most precarious conditions. For example, while all the other children in the *vecindad* (Quico, Chilindrina, and Ñoño) live at home with their parents, only (orphan and homeless) El Chavo is able to empathize with the pain of others. These timeless attributes can be seen in other fantasy characters and mythological heroes. In addition, El Chavo fails to vitiate; he never loses his innocence and his naiveté is never contaminated. Due to the "magic" of television, he remains a "child forever"; he does not age, and in this way he is instilled in the collective memory of generations, transgressing the rules of time.

However, the contradictions and ambiguities of El Chavo are not alien to viewers. The figure brings together various representations of childhood (innocence, goodness, tenderness, fragility in a context of poverty and homelessness) and tense polarities (naiveté/malice, poverty/wealth, knowledge/ignorance). This can be seen in the school scenes when Popis and Chilindrina fail to explain the discovery of America, but El Chavo is able to tell the story of Christopher Columbus and mentions that he had three "skulls"[6]: "La Santa María, La Tinta (The Ink) y La Piña (The Pineapple)". Once Profesor Jirafales gives the correct names of the three ships, the storyline continues. This comedic situation exemplifies the fact that El Chavo does not ignore historical events and that, in his own way, he is a good student. The tension between knowledge and ignorance is exposed when his character is placed on opposite poles at the same time.

El Chavo is a character traversed by ambiguity and contradiction: the former due to the excess of meanings and identifications that his character produces and the latter because the oppositional pairs that he represents offer problematic combinations. El Chavo is both good and poor, but would it be radically different if he were good and rich, or bad and poor, naive and ignorant, or naive and educated? We contend that the possible combinations and re-creations depend not only on the images offered by the program, but also on subjective productions.

To each his own Chavo

In the previous section we mentioned some of the aspects related to the figure of El Chavo, particularly how he is a complex figure generated by mass media. We now turn to discuss how the images of El Chavo act as secondary subjective productions (De Certeau, 2000). In this section, we highlight how this figure has various components whose combinations do not depend exclusively on the television, but on the subjects. By asserting that the educative occurs *through* the media, subjects have the capacity to construct a figure of El Chavo while at the same time being produced by it. This means that the production of images by subjects alters that which is offered by television and affects the configurations of our representations.

This effect is interesting because the series aired from 1971 to 1980 as an independent show and between 1981 and 1995 it became part of the "Chespirito" program, in which individual episodes were shown (Gómez Bolaños, 2007). After 1995 there were no new episodes that could incorporate more up-to-date (Mexican) sociocultural references. As such, the show's capacity to capture elements of popular culture and yet remain "unchanged" for four decades is an achievement for its creator, Roberto Gómez Bolaños. Furthermore, *El Chavo* also won the acceptance of diverse publics located at different latitudes, becoming a transgenerational figure that could affectively interpellate its viewers, provoke debate about popular culture, violence, and poverty, and—by achieving an audience that reached more than 300 million (Gómez Bolaños, 1995, p. 77)[7]—become a cultural landmark in Latin America and Europe.

Because *El Chavo del 8* has been transmitted across television screens since the 1970s (even becoming a cartoon in 2006) and has spread all over the world, we contend that the image of El Chavo has managed to remain a cultural staple because—instead of becoming petrified—it has managed to combine old, stale, ingredients with current, valid representations of the Latin American context. This can be seen in the sociopolitical and cultural processes that serve as backdrop for the series: economic poverty, inequality, abuse, and abandonment of the weakest.

While the latest generations of *El Chavo* viewers do not live or share the same history or cultural codes of the 1970s, there are some reference markers that act as an adhesive to other cultural practices, impelling the appropriation of the character. We argue that these processes are not the direct result of unidirectional television transmission (i.e., they are not simply attained or developed by being projected), but their identification is exclusively subjective.

Education *in* the media

To speak of *El Chavo del 8* is to speak of popular culture, "[which] is not the mechanical sum of the offerings of an industry, but the way in which a collective accepts and assimilates" (Monsiváis, 1981, p. 52). The following paragraphs show that what is popular about *El Chavo del 8* is not just its media production, but what society does with it.

The gap between poor and popular

When *El Chavo del 8* emerged, television was already transmitting content showcasing the plurality of urban society and its steady growth. The series' content recalls the story of the poor, the dispossessed, the marginalized and their "popular"[8] dynamics, tinged with hints of comedy. Since the 1970s, several of the series' episodes naturalize inequality, producing identificatory frames for subjects, frames that are transported via television to millions of inhabitants across Mexico and elsewhere. El Chavo's perennial desire for a "ham sandwich" is an example of the precarity and inequality that the show portrays, as none of the other characters—Señor Barriga (the owner of the housing projects), Profesor Jirafales (the school teacher), or Don Ramón (La Chilindrina's father)—can satisfy El Chavo's wish. That, is none can satiate the hunger of the poor.

The codes and symbols appropriated by large social sectors allow the characters to be placed in an urban context, highlighting their polarities: poverty and luxury, ignorance and knowledge. In the series we can see Doña Florinda as she strives to highlight the economic and social differences with Don Ramón by calling him "*chusma*" [rabble] even though the characters have more in common than she would like to acknowledge: both are single parents caring for their children, live in the same neighborhood, belong to the same social strata, wear clothing that reveals their economic deprivation, and share the same language (among other cultural codes). These tensions reflect the intercultural and political conflicts that are embodied in viewers and foster mechanisms of identification.

The series managed to disseminate a "popular language" that contained elements of comedy, the grotesque, sexual dissimulation, and double entendre. This amalgamation would become a cultural reference for the marginalized. Just like "high culture" left out the dispossessed, the ignorant, and the indigenous, this language appeared as the voice of the people, the voice of the poor, and the expelled. In this sense, El Chavo shares something with his predecessor, Cantinflas:

> Mario Moreno Cantinflas embodies the lumpen (the "peladito") . . . he provides a gait and a repertoire of gestures, he will refuse any consistency in language, and

> by acting in this way, he will be both a summary and an improvisation [...] the "cantinflesque" language where nothing is understood abounds in the streets (Monsiváis, 1978, p. 103).

A new grammar emerged, rooted in expressions taken from urban contexts that transcended the territories demarcated by the city. That language of the streets comes into the *vecindad* and is showcased as the voice of the marginalized, the *chusma*. Throughout the series, phrases of popular jargon are constantly repeated: "*qué burro*" [what an arse], "*si serás*" [of course you are!], "*fíjate, fíjate, fíjate*" [look, look, look]. And, at the same time, the characters of the series add to this particular language with phrases like: "*se me chispoteó*" [this untranslatable made-up phrase refers to situations in which impertinent or neglected ideas are blurted out], "*me lleva el chanfle*" [another made-up phrase similar to "Damn!"], "*fue sin querer queriendo*" [I did it without meaning to, by accident], and "*tenía que ser El chavo del 8*" [it had to be El Chavo!].

What a nice *vecindad*!

El Chavo del 8's success of more than forty-five years can be explained if we consider it as a media product that includes various elements of popular culture and which becomes a frame of reference available for subjects to appropriate. In the following paragraphs, we discuss how the television series becomes an educative media process through the *vecindad*, and from there, projects an aesthetic experience.[9]

From the fields to the *vecindad*

In the 1930s, Mexico City experienced a large migratory wave as several social sectors were "expelled" from the Mexican countryside and took up residence in the central areas of the capital. Such movement was due to changes in the forms of production in the Mexican countryside, changes in state policies directed towards rural sectors, the gradual abandonment of land reform proposals which were a legacy of the revolutionary movement of 1910, and a lack of investment in communal production models (along with their gradual privatization) in the northern part of the country (Carton, 2009).

The industrialization process that took place in Mexico in the 1940s changed urban life. The City of Palaces—a name given to Mexico City due its French-inspired nineteenth-century architecture—for example, was constantly transforming. Starting in the 1930s, the expansion of the capital city began. "The goal was to give functionality to the urban space, make it suitable for future growth of the metropolis and allow heavy investment in 'urban' renovation" (Sánchez-Mejorada, 2005, pp. 280–282 as cited by Quiroz, 2013). In 1930, Mexico City had just over 1 million people; by 1950 that number was close to 3 million (Quiroz, 2013). By 1970, 6,874,165 inhabitants were reported (IX General Population Census, 1970).

The occupation of these bordering territories by industry made it difficult to find housing spaces; in order to protect the integrity of the family, workers of different backgrounds and economic conditions settled in whichever spaces were available: people settled in small plots of land with wooden buildings, in warehouses, abandoned or demolished houses, and large residential homes, among others. In such a context, the *vecindad*, also known as a *casa de patios* [a house of courtyards] or *conventillos* [tenements] became "popular" housing.

Although the origins of the *vecindad* stretch back to the nineteenth century, it became an established part of the urban landscape by the middle of the twentieth century, generating particular life conditions. A *vecindad* is characterized by overcrowding, poverty, and social inequality. While its inhabitants have a range of occupations and professions, most subsist on precarious incomes. The *vecindad* is an area (usually an old house with hallways and wide corridors) with an interior that is often distributed in the following way: a set of small houses or apartments organized around an interior courtyard which becomes the centralized "common" space (a dynamic that is visible in *El Chavo del 8*), and where bathrooms are typically shared and located next to the laundry facilities, another common area.[10]

The *vecindad*, the marginal in the heart of the city

Life in the *vecindad* implies a political exercise that involves transgressing time. By 1960 provisions and regulations relocated many of these housing spaces to plots surrounding the city center. There was not a single government representative who did not want to "erase" a number of these *vecindades* from the map, a wish that showcased the political status of these housing sectors, stigmatized as the nucleus of poverty and inequality. They represented a large vulnerable social sector (also called the "marginal") and fought against housing redistribution policies.

The resilience of the *vecindad* and its inhabitants helped them remain in the economic and administrative center of Mexico City, placing "marginality at the core" (Monsiváis, 2000). That is, their presence emphasized poverty, overcrowding, and social inequality in the heart of the capital.[11] In this regard we contend that *El Chavo del 8* has kept that experience alive, transgressing temporality and spatiality. The media and educational effect achieved is relevant because the nineteenth-century *vecindad* remains entrenched in popular culture as a symbol of poverty and is part of the city's fabric. Thus, *El Chavo del 8* contributes to the imaginary construction of the city; the show "draws" the city while at the same time is represented by it.

One of the striking features of the series is its ability to recreate, transmit, and share such experiences: to live in the *vecindad* of El Chavo is to inhabit the city. Although the series only shows some of the houses of the *vecindad*, it is understood that the city is "there." According to Silva (1993):

> The doorway of the house, the threshold to cross from the private world to the public of the city, could be that disturbing border that summarizes the two key territories that have to accompany us throughout life: being inside (the house) and being outside (in the city) (p. 19).

When *El Chavo del 8* appears on television, there are whole generations who share meanings about the practices of coexistence in the *vecindad*. In several episodes we see the characters as they go through doorways into community life; the city shares the same name as the street and references are made to a restaurant, a school, in an effort to name the "outside," an outside that is essential so as to think about life *inside* the *vecindad*. In many of the episodes the plot unfolds in the central courtyard surrounded by the various apartments; however, sometimes we see the backdrop of Doña Florinda's apartment or the classroom of a school. Although these backdrops are simple and sparse due to the lack of scenic resources, it is important to think about how the "intimacy" of the *vecindad* was also built *sin querer queriendo*.

The aesthetic experience that links the intimate life of a *vecindad* has roots in ancient times because indigenous residential complexes had a similar organization for their neighborhoods. These communities had a *Cihuacalli*, which represented a meeting place, a space for sharing, welcoming, and receiving (López, 1989). To live in such a neighborhood was akin to living in a family provincial home, with collective dynamics that involved the distribution of work, income, social interaction, and so on.[12]

The *vecindad* is the city, the barrel is the house

To think of El Chavo as an inhabitant of the *vecindad* involves first recognizing him as the subject of an aesthetic experience. Anecdotes of El Chavo appeal, at least in part, to the memories, emotions, interactions, and symbolic exchanges that occur in the neighborhood. To watch El Chavo is to look at oneself, not only as a mechanical reflex or as a reproduction of the media's content, but to feel part of a common history, a collective *pathos*. The media education that *El Chavo del 8* provides involves identifying the different ways of inhabiting the *vecindad*.

In this sense, the *vecindad* can be considered an urban space with the capacity to symbolize habitat. It represents a social nucleus; a territorial and symbolic space that offers markers of identification: the neighborhood as a device that orders individual and collective daily practices; as a physical territory that delineates a context; as a symbolic space that generates meanings about one's being in the world. It even has a pedagogy of living that involves an aesthetic experience: the subject inhabits a common space, its corporeality crossed by other bodies, its senses and meanings constructed in relation to these spaces:

> Human bodies and the body of a city continue to complement each other to the extent that the former occupies the latter and in that act of occupation learns not only to live its air but to dream with what the city can offer [. . .] To cross its limits can lead to the punishment of those who live in a forbidden space, or it can also lead to an act of creation and discovery from the other side of the city. Thus, the city always seems a like body, waiting to be discovered (Silva, 1993, p. 22).

This can be seen in the series when playing with a ball or hopscotch takes place in homes characterized by poverty, inequality, and overcrowding. As already mentioned, the *vecindad* symbolically represents the city, so to play there is a way to enjoy the city. To play hopscotch in the *vecindad* does not change the rules of the game, but rather the relationships of the game. In El Chavo's *vecindad*, one does not have to look for a partner to play; the presence of the *vecindad's* inhabitants provides the opportunities for such play.

The *vecindad* fosters relationships between its inhabitants: the disputes between La Chilindrina and Quico generate disputes between Don Ramón (father of the first) and Doña Florinda (mother of the second), be it over a cart or a doll. The closeness and contact between the characters is a result of the sharing of the collective; the *vecindad* conditions spaces and bodies.

These provisions allow the *vecindad* to be presented as an analogy of the city and its inhabitants as the metaphoric city dwellers. The aesthetic experience that the *vecindad* offers has a collective and political dimension with which the viewer can identify: the neighborhood, the courtyard, the local family restaurant or tavern. These elements bind and retain lost worlds, collective and private. In the show we can see characters in their homes, such as Doña Clotilde, the "Witch of Apt. # 71."[13]

The characters and their interactions show the *vecindad* in its natural state: where lack of safety, inequality, and poverty are part of the habitat. An example of this can be seen when El Chavo takes refuge in a barrel located in the central courtyard, a space that belongs to all of the *vecindad's* inhabitants. The series helps promote the growing perception that El Chavo is not only part of the neighborhood's landscape, he is also the neighborhood itself.

We conclude this section by outlining the particular (and collective) dynamics of interaction in El Chavo's *vecindad* that give meaning to the aesthetic experience:

1. The *vecindad lives*: it is a vital living space. Here, territories are demarcated, experiences are distributed, and spaces are shared. Many of the events are held in the central courtyard, marking it as a meeting point of different cultures and bodies.
2. The *vecindad feels*: there is an emotional charge linked to the space. For example, the presence of Profesor Jirafales in the neighborhood is due to his ongoing love affair with Doña Florinda, with the central courtyard acting as a witness to their courtship. The affective charge consigned to the *vecindad* is reinforced by its inhabitants' interactions. This space is symbolized by the affections, both its associations and aversions, of those who live there.
3. The *vecindad* is a symbolic territory: although the series itself does not provide a reference to its location in the city, the viewers are located beyond its physical territory.

As such, we argue that the television series produces an aesthetic experience that is also part of its media education. To go to the *vecindad* is not just a way to reiterate a cliché of poverty in the entertainment industry, it also offers the possibility of recreating or transforming an historical reference that is continuously actualized through the television screen and is articulated in the daily lives of its viewers.

Although *El Chavo del 8* does not incite social criticism, one could argue that the show had some sensitivity in trying to capture and recreate the daily conditions of a particular era. In the series, viewers of different social strata are exposed to a precarious *vecindad*. The show's scripts produce a reality for viewers of other urban or rural environments: the city is right "there," close and asymmetrical.

The educative *against* the media

In the previous sections we explored some dimensions of media education, including how an aesthetic experience can evoke a relationship with a city, as well as life in a neighborhood that emphasizes certain community features. These elements, in our view, shape cultural reference frames, which then modulate our own representations. Additionally, we have also argued that media education occurs *through* the media: by creating links (albeit contingent ones) between various elements, some of which are derived from the cultural reference frames provided by television, and others are drawn from symbolic and cultural records that not only respond to the cultural mainstream but to the production of subjectivity as well.

Now, we want to stress another productive and political facet: education *against* the media. This refers to a production of subjects which subvert the "intended" order of *mass media*, the cultural *mainstream* or the dominant media discourses. Although this should not be understood as the subject's will to resist the lure of television, it is important to note that what is educational about going *against* the media is that it affords the possibility of producing "something more" than identifications. In this regard, we return to a criticism made by de Certeau on Foucault's panopticon as a power-monitoring device on bodies and spaces. According to de Certeau (2000):

> If it is true that everywhere the extending "grid" of discipline is becoming clearer and more extensive, it is all the more urgent to discover how an entire society resists being reduced to it, what popular procedures (also "miniscule" and quotidian) manipulate the mechanisms of discipline and conform to them only in order to evade them, and finally, what "ways of operating" form the counterpart, on the consumer's (or "dominee's"?) side, of the mute processes that organize the establishment of socioeconomic order (p. 44).

In this vein, de Certeau (2000) helps us to see how television as a device analogous to how a panopticon operates, not only by producing identifications among viewers according to some ideological, economic, and social content, but also by encouraging other practices that de Certeau (2000) calls "antidisciplinary" (p. 45). From our viewpoint, these practices can be isolated and precarious; they are neither intentional nor strategic, but go against the dominant cultural patterns. Their function is to show that, from the other side of the screen, content other than that launched by the media, can occur.

Mass media against mass media

While it now seems commonsensical to say that television leads to a strengthening of particular ideological content, this was not always the case. The introduction of television in Mexico operated against other media technology such as film and radio, both of which were hegemonic during the last decades of the nineteenth century and the early part of the twentieth century:

> [In the early part of the twentieth century] viewers are discovered and recognized thanks to their scenic counterparts [...] *The Stock character*: men with huge mustaches, drunks, scorned lovers, ranchers longing for refinement [...] This public, thanks to the [Mexican Revolution of 1910] begins to solidify itself based on its contemplation of the scene [...] this first urban popular culture was a weapon that was armed by a nationalist decision: we can count on this, this is who we are, and this is how we speak (Monsiváis, 1978, p. 100).

We can think of other cultural devices under the same lens: Mexican mural painting, comics and leaflets portraying urban life, as well as cinema and radio between 1930 and 1950. These "mass media" expanded access to culture in the Mexican context and were pioneers in proposing media education.

In the middle of the twentieth century, urban life demonstrated a dynamic cultural mosaic that would gradually and eventually settle into a symbolic machinery. Daily life marked the rhythm of cultural interactions with "[...] 1950 [as] the defining date. TV gets to Mexico; the rules of the game change, encouraging nationalist dominance and a manipulative industrial colonization take charge of the forms and methods of popular urban culture" (Monsiváis, 1978, p. 99). Perhaps in the 1970s it was thought that producing a television program would captivate the public without reflecting on the impact that such projection (of poverty on the small screen) would have on popular culture. But, even in the last decades of the twentieth century, *El Chavo del 8* continued to use a "particular" vernacular language that acted as social glue, becoming "popular" through its scope and content.

Within this logic, the media education that the series offers is not simple transmission; there is an education that, in the case of television, is technological.

The transmission of the series for over forty years has not only contributed to the formation of subjects and their imbrication with urban popular culture, it has also disrupted the chronological laws to represent a past-present: a continuum that presents the old and updates the past.

The marginal as a lifestyle

El Chavo del 8 showcases the marginal as a way of life by offering cultural frames of identification. Although they share poverty and overcrowding, the characters of the *vecindad* accentuate their social differences. Through the figure of the "marginalized" or the *chusma* [rabble], *El Chavo del 8* strengthened the stereotype of the poor by showcasing their precariousness while at the same time fostering commiseration:

> [The child that inspired El Chavo] wore a pair of "miner" shoes that had obviously belonged to an adult. But perhaps his most characteristic piece of clothing was his old hat with earflaps, which in cold weather might have served some use, but when I met him, in the summer, did nothing but accentuate the grotesqueness of his figure (Gómez Bolaños, 1995, p 2).

Through this image, the series showcased the community ties of those assumed to be the *chusma*. It showed a group of poor people who shared the culture of the *vecindad*, a set of linguistic codes, patterns of behavior and recognition, while at the same time fencing it off from the rest of the city.

To mock poverty on television was a way to exorcise it. Even though life in the *vecindad* was shown on television for domestic consumption, the series exported these images to different countries and received a tremendous reception among diverse social sectors. Through the show, the people of Brazilian *favelas*, Argentine tenements, or the Mexican *barrio* of Tepito, found a space where, far from being sanctioned, they were sheltered with affection and acceptance.

The degree of incorporation of these frames of reference has transcended the television series. Current generations adopt the forms and linguistic codes that were once considered marginal. This can be observed in the costume design of Don Ramón, El Chapulín Colorado (another character created by Gómez Bolaños), Cantinflas (a Mexican comic film actor), or El Santo (a Mexican wrestler). The repetition of clichés also helps anchor notions of poverty and inequality, but that is not all that is achieved through media education. Values, feelings, patterns of identification which connect subjects and affect the construction of ethical-political sites other than those that occurred during the 1970s, are also produced. By not ignoring the importance of the effects of televised content and practices on the cultural mainstream, it is important to think about what *El Chavo del 8* transmits to audiences today.

Final thoughts: "*sin querer queriendo*"

This chapter has allowed us to sharpen our gaze on a media phenomenon that continues to provide dimensions for analysis, ranging from rhetoric to semiotics to linguistics, among others. The view that we developed supports the idea that *El Chavo del 8* is a cultural product with a strong educational element that cannot be reduced to the consumption of content transmitted by television itself. In this regard, and as a final reflection, we would like to make some clarifications.

To affirm that education happens *through* the media alludes to the *intermezzo*, the space between the televised content and the identification of its subjects. It represents an intermediary and intermediate zone, a space in which the image projected by the mass media creates the conditions of (im)possibility for the meanings and elements of various sources to emerge as sites for identification. The educational element happens when the image of *El Chavo del 8* raises diverse meanings in distant contexts across different generations, transgressing the laws of time and space to reach our screens as a present reminder, like a hologram that makes possible our *chavo*.

We also contend that education occurs *in* the media. In our analysis we consider television as a site that transmits and proposes certain cultural reference frames that are required to produce identification. We emphasize the aesthetic character of the identifications fostered by the representation of the *vecindad* as habitat. Education in the media refers not only to television, but also to the "medium" in which the characters interact. The *vecindad* is a medium in which both the bodies and their stories can be encountered. The series brings the dynamics of the city to the *vecindad*, converting it into the city; an aesthetic experience that occurs through media education.

Although mass media provides cultural references that frame certain identifications, we know that educative processes often bring together elements from different origins that are not necessarily intended by the media. Education *against* the media also accentuates its unforeseen effects on subjectivity, as exemplified by the series' ability to become a representative figure of the marginal, loaded with both emotional and political elements.

Finally, it is important to recognize that these ideas have been generated from the symbolic, emotional, affective, and intercultural reconstructions evoked by the figure and face of *El Chavo del 8*. Our analysis is mixed with the inheritances that the series bequeathed to us through its characters in Mexico and other parts of Latin America. Woven with feelings that were (and continue to be) superficial inscriptions of "other's faces," we want to emphasize that the series showcases images of everyday urban life, images marked by the peculiarities, humor, platitudes, simplicity, and exclusions of life in the *vecindad*. Simple and complex

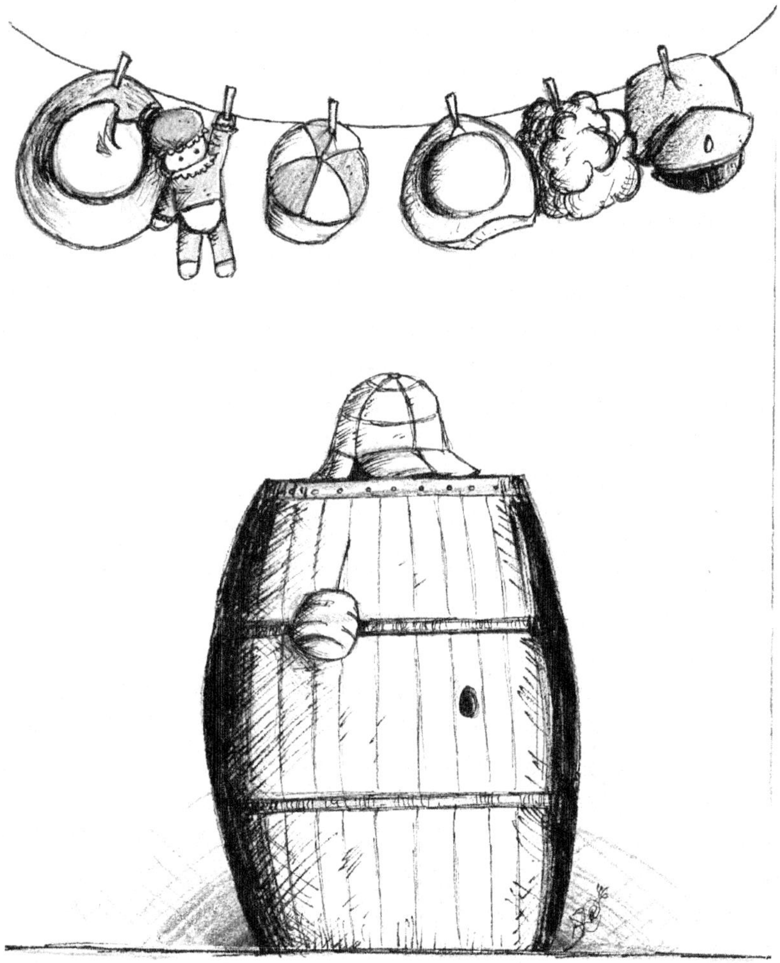

This picture is an artistic creation of Dorian Flores Reyes, whom we would like to thank.

at the same time, the aesthetic, political, and comedic dimensions of the series helped it emerge as an educational site *sin querer queriendo.*

Notes

1 These housing projects, all built on the same plot of land, were occupied by different families.

2 Editors' note: many of these phrases were invented by the show and are almost impossible to translate. Please see the glossary at the end of the book for a more extensive explanation of each one.

3 This refers to a political component, to the extent that education involves an exercise of power that alters the relationship(s) and interaction(s) between the elements involved.

4 Nor could one foresee the transition from television to other forms of digital media, or the convergence of this comedy of errors becoming a cartoon in 2006. Since 2006, *El Chavo del 8* has reproduced the original characters and episodes as a cartoon.

5 These iterations can be considered as a process of difference in repetition (Derrida, 1998).

6 Rather than name the three *carabelas*, or Spanish galleons, El Chavo mistakenly uses the similar word, *calaveras,* which means "skulls." He also gives the incorrect name for two of Columbus's ships, stating "La Tinta" and "La Piña" instead of "La Niña" and "La Pinta."

7 These figures are not official, but in 2012, Forbes affirmed that 91 million people watched *El Chavo del 8* on a daily basis, while Univision reports that, at first, the show reached a rating of 60 points. In 2011, Excélsior claimed that, in 1975, the series was watched by 350 million people a week (Resendiz, 2014).

8 Translator's note: The term *popular* in Spanish connotes a double meaning of something that has spread among a large portion of the population, and of something that is specific to the lower classes. This ambiguity between the whole and a part is wonderfully explored by Giorgio Agamben (2000) in his essay *What is a People?*

9 To inhabit the *vecindad* represents an aesthetic experience to the extent that it floods the senses, is rooted in (and exceeds) culture, and involves modes of existence in a space that is not limited to corporeality. To live is to take a space, make it one's own, and this appropriation is a unique manifestation of existence. This manner of inscribing oneself in a space and creating a way of life is highly political to the extent that it subverts the established order and promotes subjective instances in the "social order."

10 "[...] On average there are 2.87 people per room, 25.76 people per shower/bath, 10.43 per toilet, and 8.11 per laundry facility. However, there were plenty of extreme cases, such as house # 8 in *vecindad* #72 on Rosario Street in a central area of the city. This house consisted of a room 3 by 4 meters wide and 4 meters high. Eleven people lived in this space: a couple and their nine children; with only two beds to share among all of them" (Quiroz, 2013, p. 35).

11 Although in the twenty-first century *vecindades* have become architectural heritage sites, they no longer have any inhabitants. They have been replaced by multi-family buildings.

12 These family practices ranged from the distribution of economic and domestic work between parents and children to the political representation of entire groups designated by a common name.

13 Ms Clotilde is an elderly woman who lives in apartment 71.

References

Butler, J. (2010). *Marcos de guerra. Las vidas lloradas*. México: Paidós.

Carbajal, J. (2003). "Internet, lo educativo y la educación: complejo discursivo." In: Granja, J. (ed.), *Miradas a lo educativo Exploraciones en los límites*. México: SADE-Plaza y Valdez, pp. 43–64.

Carton de Grammont, H. (2009). La desagrarización del campo mexicano. *Convergencia, 16*(50), 13–55. Retrieved April 9, 2016, from www.scielo.org.mx/scielo.php?script=sci_arttext&pid=S1405-14352009000200002&lng=es&tlng=es.

De Certeau, M. (2000). *La invención de lo cotidiano I. Artes de hacer*. México: ITESO-UIA.

Deleuze, G. & Guattari, F. (2006). *Mil mesetas. Capitalismo y Esquizofrenia*. Spain: Pretextos.

Derrida, J. (1998). *Márgenes de la Filosofía*. Madrid: Cátedra.

Eco, H. (1984). *Apocalípticos e integrados*. Spain: Lumen.

Gómez Bolaños, R. (2007). *Sin querer queriendo Memorias*. México: Aguilar.

Gómez Bolaños, R. (1995). *El diario del Chavo del 8*. México: G punto de lectura.

Gómez Bolaños, R. (1979). "Qué bonita vecindad." In: *Así Cantamos y Vacilamos en la Vecindad del Chavo*. México: Phonogram. Retrieved from: www.chavodel8.com/cancion/letra-que-bonita-vecindad/.

Instituto Nacional de Geografía e Historia. IX Censo General de Población de 1970. México: Inegi. Retrieved on May 15, 2016, from: www3.inegi.org.mx/sistemas/tabuladosbasicos/tabdirecto.aspx?s=est&c=16763.

López, F. (1989). *Arquitectura vernácula*. México: Trillas.

McLuhan, M. & Powers, B. (1995). *La aldea global. Transformaciones en la vida y los medios de comunicación mundiales en el siglo XXI*. Barcelona: Gedisa.

Monsiváis, C. (2000). *Salvador Novo: Lo marginal en el centro*. México: Era.

Monsiváis, C. (1981). "Notas sobre el Estado, la cultura nacional y las culturas populares en México." In: *Cuadernos Políticos*, (30), México: Era, pp. 33–52.

Monsiváis, C. (1978). "Notas sobre la cultura popular en México." En *Latin American Perspectives*, 5, (1), (Winter), pp. 98–118.

Quiroz, M. (2013). "Las vecindades en la Ciudad de México. Un problema de modernidad, 1940–1952." In: *Historia 2.0 Conocimiento Histórico en clave digital*. III, (6), pp. 28–42.

Reséndiz, L. (2014). "¿Cómo explicar el éxito de Chespirito?" In: *Letras Libres*, (2), 2014. Consultado el día 20 de mayo de 2016. En: www.letraslibres.com/blogs/en-pantalla/como-explicar-el-exito-de-chespirito.

Sartori, G. (1998). *Homo videns la sociedad teledirigida*. Argentina: Taurus.

Silva, A. (1993). "Ciudad imaginada." In: *Signo y pensamiento*. 22. Colombia: Pontificia Universidad Javeriana, pp. 19–28.

Coda

El Chavo del 8's Connections and Reverberations

Erica Colmenares and Daniel Friedrich

How does one conclude a book that eschews fixed and stable meanings, whose attention wanders in multiple directions, and whose expressions are indeterminate, indefinite, and provisional? Is such a task even possible, or do we face the Sisyphean prospect of pushing a giant intellectual boulder up a hill, only to watch our work fall back down, forcing us to repeat our efforts again? A volume like this operates like a rhizome; it is intended to be read not as a box with something inside it that one tries to decipher and decode, but rather as a non-signifying machine that one reads "intensively," figuring how the text works and what it might connect to (Deleuze and Guattari, 1987). Consequently, rather than bring this book to some artificial close with clear lessons to be learned and steps to be taken, in the spirit of a musical *da capo al coda*—where a musician returns to the beginning of the piece before he/she reaches the *coda*, or tail(end)—we start this chapter with a series of connections that sparked, or coincided, with the creation of this rhizome book. We follow that with some questions and then offer the reader multiple *trajets*, or paths, that they might choose to follow.

Da Capo

Connection # 1

October 2015, an ordinary conversation between friends and co-editors, advisor (Dani Friedrich) and advisee (Erica Colmenares). During a weekly dissertation "check in" meeting, Erica looks over to Dani's meticulously arranged office library and notices a new figurine nestled between his books, a plush *El Chavo* toy. The sight derails Erica's train of thought. "Ay, El Chavooooo!" she coos. Erica clutches the figurine and nuzzles it against her cheek. "God, I loved that show. I would watch it every day after school." She rubs the toy's head, and thinks about her *El Chavo* refrigerator magnet collection and how only a few months earlier (as a 31-year-old-adult) she had spent many of her summer months eating at different McDonald's® restaurants in her home country of Venezuela with the hope of obtaining the entire *El Chavo* Happy Meal® toy collection. But, instead of revealing these details, Erica places the toy back on its perch, sighs, and returns to her dissertation notebook. Unexpectedly, Dani leans backwards in his office chair and nonchalantly remarks, "You know, one day I am going to write a book about *El Chavo*." Erica chuckles. While she finds the prospect profoundly exhilarating, it is also brazen and incredulous. After all, in the era of total bureaucratization of the U.S. academy, where research that is not clearly practical, marketable, and profitable is often deemed unworthy (Graeber, 2015),[1] who would ever publish a book on *El Chavo del 8*, let alone in *English*? In spite of these challenges, however, we, Dani and Erica, drafted a book proposal and sent it off. A few months later, we were elated when it was approved. All of a sudden, here we were, two native Spanish speakers from opposite ends of the South American continent—Argentina and Venezuela—tasked with writing about *El Chavo del 8*, a Mexican comedy series, to a publishing company on the other side of the ocean (the UK), in *English*. And yet, to *not* write about it would be unthinkable.

Connection #2

November 2015, a mourning son looking for lightness. Two months earlier, in September 2015, Dani's beloved father, Beno, had passed away at the age of 62 after a long and hard battle with cancer. The battle had taken its toll, and Dani found it almost impossible to reconnect with his work and get past the blank page. But how does one write about democratic education, privatizing trends, neoliberal globalization, or teacher education reform when it is so hard to get out of bed in the morning? Yet, one afternoon, coming back home, Dani hears the familiar "*¡Fue sin querer queriendo!*" [I did it without meaning to] followed by the even more familiar laughter coming from an unexpected place. Javi, the middle-aged Dominican doorman, was watching *El Chavo* on his cellphone. Dani ran back to his apartment and looked to stream the show next to his four-year-old son, Liam. Despite the language barrier, the giggles started as soon as the character Quico began his famous cry. Javi, Liam, echoes of a young Dani sitting next to Beno, a seemingly light moment that infused laughter and pleasure back into the everyday. To *not* write about El Chavo would be unthinkable.

Mutliplicitous connections

Encounters and connections to *El Chavo* seemed to occur *en todos lados, todo el tiempo* [everywhere, all the time]: The mention of an *El Chavo* catchphrase at an academic conference in the U.S. that provokes wide-eyed recognition and laughter by those in the know; an old, broken-down toy at a flea market evoking memories of life before on-demand media, where you *had* to run home to watch a show; a young boy and his mother giggling while watching an *El Chavo* episode on a cellphone on a crowded NYC subway; a Halloween costume made of a red unitard, a yellow heart logo with "CH" in the middle and bouncy antennae[2] that expose a cultural divide; the shock of bumping into a costumed El Chavo among the Naked Cowboys, Lady Liberties, Power Rangers, and Elmos of Times Square; watching media(ted) images of schooling for the poor and the wealthy, sitting together under the same roof, loved by a teacher concerned about their future, juxtaposed with our experiences of New York City's public schools, as segregated as they have always been; and noting the palpable contrast between the children of the *vecindad,* left alone to play for hours without surveillance and constant guardianship, among numerous others. The bombardment of uncanny and (laughably) prophetic encounters and connections to *El Chavo* further strengthened our desire to write about the show.

Diving into middles, finding mediators, and creating plateaus

With the book proposal approved, we dove headfirst into the middle, for "things and thoughts advance or grow out from the middle . . . and that's where you have to get to work, that's where everything unfolds" (Deleuze, 1995, pp. 160–161). We did not have a map, scheme, or model to guide and structure our thoughts. Inspired by the work of Deleuze and Guattari (1987), we were more interested in the notion of experimenting with ideas, opening new possibilities, and making multiple connections.[3] Our aim was not to judge *El Chavo del 8* (our nostalgic entanglements with the series are impossible to escape), offer some clever analysis of the show, or subject it to further "interpretotis" (Deleuze and Guattari, 1987, p. 114), all of which presume to know what is to come. Instead, we wanted to remain "attentive to the unknown knocking at the door" (Rajchman, 2000, p. 7), even if that meant being unaware of what could happen.

But, we knew we could not do this work alone. To move from thinking about the multiple "to actually *making* the multiple" (Tomlinson and Burchell,

1994, p. viii, our emphasis) we sought collaboration with others. In our search, we aimed to avoid what Deleuze (1995) calls "imitators," likeminded others whose primary interest is to mimic and forge copies instead of building new creations. So, we invited people both in and outside our fields of curriculum studies and teacher education, and looked for individuals connected to different interests, from diverse backgrounds, with assorted experiences. Our goal was to convene a group of "mediators," or intercessors (Deleuze, 1995), who could help us think—and therefore create—new ideas about *El Chavo del 8*. According to Deleuze (1995):

> Mediators are fundamental. Creation's all about mediators. Without them, nothing happens. Whether they're real or imaginary, animate or inanimate, you have to form your mediators. It's a series. If you're not in some series, even a completely imaginary one, you're lost (p. 125).

Our mediators—Nicolás Arata, Victoria Parra-Moreno, Ana Paula Marques de Carvalho, Rita Frangella, Carlos Aguasaco, Limarys Caraballo, Ernesto Treviño Ronzón, Dulce María Cabrera, and José Carbajal Romero—proved invaluable. They helped make visible what might have otherwise remained hidden (Deleuze, 1995). Like Jorge Luis Borges' labyrinth of ever-bifurcating paths, they helped us forge inroads into "complex *themes*" (Rajchman, 2000) such as affect (see Chapter 6), education and schooling (see Chapters 1 and 3), curriculum studies (see Chapters 1 and 2), childhood(s) (see Chapters 2 and 3), media (see Chapters 7 and 8), popular culture (see Chapters 1 and 5), literacy (see Chapter 4), and cultural studies (see Chapter 8); to a variety of *scholars:* Arjun Appadurai, Roland Barthes, Homi Bhabha, Gilles Deleuze, Jacques Derrida, Michel Foucault, Felix Guattari, Jacques Rancière, Slavoj Žižek, among others; and *objects*—Pepito cartoons, tenement buildings, television sets, and seemingly a-temporal classrooms.

When it came time to arrange these unique connections, we created an assemblage of three non-sequential, interrelated plateaus—Plateau I: El Chavo Encounters Latin American Education and Childhoods, Plateau II: El Chavo's Encounters with other Latin American Societies and Cultural Artifacts, and Plateau III: El Chavo's *Media*-ted Encounters. Although this organization might imply some teleology, any overarching meaning, purpose, or progress is unintentional. Instead of reaching a climax or (end)point, the plateaus are nonhierarchal zones of sustained intensity intended to leave the reader with "an afterimage of its dynamism" (Massumi, 1992, p. 7) so that the energy might be harnessed and reinserted elsewhere, making way for new trajectories and movements (Waterhouse, 2012). Thus, each plateau is a multiplicity connected to other multiplicities (i.e. plateaus) to form a rhizome of endless connections.

Working with Deleuze and Guattari's (1987) principle of "disjunctive synthesis" or "a relation of non-relation," the book holds together disparate elements without abolishing what makes them different. As the reader might have already noticed, contradictions and tensions are left unresolved and all "circles of convergence" (Deleuze and Guattari, 1989, p. 22) are rooted in difference, not sameness. Like the musical *ritornello* that works through difference and complex repetition (e.g. repetition of difference versus repetition of the same), we embraced this difference for it is only through it that something new about *El Chavo del 8* can be created.

Here, let us highlight some of the differences, as we hope the friction between the ideas and approaches presented by the various authors might generate catalysts for future engagements and connections. Perhaps the most salient of these is between the tone of Treviño Ronzón's (Chapter 7) and Cabrera and Carbajal's (Chapter 8) texts and the rest of the book. These two chapters take on a much more aggressively critical stance towards Roberto Gómez Bolaños's creations than the other chapters, making arguments about the show's complicity with corrupt political regimes and the media conglomerate's (i.e. Televisa) deep ties to political figures of the time. While Colmenares (Chapter 6) and Caraballo (Chapter 5) wax poetic about *El Chavo's* affective engagements with their own childhoods, Treviño Ronzón charges against the romantization of poverty as a mediated space that tends to foreclose a more critical view towards a country on the verge of collapse. We find it worth noting that Treviño Ronzón, Carbajal, and Cabrera are the only authors who grew up in the Mexico that gave birth to *El Chavo*, revealing what several scholars we talked to had warned us about: the conflicted relationship between Chespirito and his birthplace. The other authors, including the book's editors, experienced Gómez Bolaños (and his characters) as an "indigenous foreigner" (Popkewitz, 2005), a reinvention of the characters in places far removed from Televisa's reign and usually through reruns viewed decades after the massacre of Tlatelolco.[4] While we obviously do not want to imply that one cannot be critical towards objects of affection, this is only to note that what may seem like a Latin American dialogue is sometimes a melange of discursive lines of flight, intersecting only in particular points.

Beyond the difference in tone linked to different intensities and varying affective entanglements, there is a question that lingers through the book that each chapter answers differently. Is *El Chavo* a revolutionary interruption of traditional notions of childhood, schooling, and society, one that is presented as an alternative to the disintegrating imaginaries of progress and the good life? Or is the show a sophisticated attempt at legitimating symbolic and physical violence, an endeavor to pass extreme poverty as funny, a fable to instill the lesson that all that is needed to fix society's problems is a good hug or a community barbecue by the beach? As we hope every single author in this volume has made clear, these oppositional binaries are not useful when thinking through Chespirito's creation(s) and its effect on Latin American societies. This book works with the logic of AND (over OR): *El Chavo* is *all* of those things and many more, which is precisely what makes this fictional character such fertile ground for continuous entanglements, engagements, and connections.

In this sense, we might say that this rhizome book has parallels with what Deleuze and Guattari (1986) refer to as "minor literature," a kind of literature that opens up new perspectives, paves the way for new avenues of thought, and breaks from established forms. Deleuze and Guattari credit Franz Kafka[5] with initiating this form of literature; they often describe minor literature by contrasting it with "major literature." Unlike major literature that is weighted down by old categories of genres, models, and styles that require readers to interpret, minor literature is a *reversal*, a "*rallying point*" (Bensmaïa, 1986, p. xiv, original emphasis) for texts and reading and writing practices that are often marginalized or ignored. Like this rhizome book, minor literature is not intent on expressing a particular meaning, representing a specific thing or being, or imitating some precise nature (Bensmaïa, 1986). Nor does it privilege an individual subject (i.e. author). In contrast to major literature that is primarily concerned with the individuated enunciation of some literary master, minor literature expresses itself in collective terms and produces collective utterances (Bensmaïa, 1986; Deleuze, 1989). This book works in the same way: to attribute what was created here to any particular author is to ignore the working of "variously formed matters . . . speeds" (Deleuze and Parnet, 2007, p. 3) and risks the indignity of speaking for others (Foucault and Deleuze, 1977). Fundamentally, this text is a collective assemblage of enunciation (Deleuze, 1989).

Furthermore, minor literature is often born out of absolute necessity—in restricted conditions and out of a particular *milieu* (Deleuze, 1997)—that habitually breaks from established forms, encourages new directions, and releases particular potentials (Deleuze and Guattari, 1986; Rajchman, 2000). Without falling into explication or attributing causality, is it possible to extend this idea of minor literature not just to this book itself, but to *El Chavo del 8*, as well? After all, the *milieu* in which *El Chavo del 8* emerged was unique. Politically, Latin America was experiencing one of its most unsettling decades: locked in a cycle of authoritarian regimes, bloody coups, guerrilla struggles, civil conflicts, and foreign intervention, there was a mass migration of people within and across borders, a rapid process of urbanization that swept through the continent's major cities, a growing Americanization of global popular culture and consumption, and widely shifting economies. Although *El Chavo del 8* was a television series intended for the masses and replete with clichés,[6] we want to suggest that it took up a method of "writing, creating, and making" that was perhaps rooted in a minor literature (Deleuze and Guattari, 1986). In a *milieu* long dominated by U.S. sitcoms dubbed over with the same Spanish voices, *El Chavo del 8,* like minor literature, "picked up" and "stole" from here and there, deterritorializing hegemonic television practices and created something that Latin America had likely never seen or experienced before. Despite a precariously limited budget during its initial years, the show itself became a site of active experimentation: it relied on double entendres and misinterpretation, used adult actors to play the role of children, and exposed the continent to Spanish words and phrases that had never before been heard outside of particular parts of Mexico. Given these contingencies and circumstances, can we consider *El Chavo* as "a visionary phenomenon, as if society suddenly saw what was intolerable in it and also saw the possibility for something else" (Deleuze, 2006, p. 234)? Is it a "collective and political phenomenon in the form of: 'Give me the possible, or else I'll suffocate'" ... (Deleuze, 2006, p. 234)? We leave that question for you to ponder.

Although *El Chavo del 8* lives on through reruns, merchandise, and cartoon spinoffs, the show has not been in production for over a quarter of a century. While younger generations across Latin America are likely to have some familiarity with the show—perhaps they have heard about it from older generations, have seen reruns, bumped into El Chavo merchandise in stores, or currently watch the contemporary El Chavo cartoon spinoff—they did not viscerally live or experience the show like those of us glued to our television sets sometime between the 1970s to 1990s. This is not to suggest that every Latin American during that time period watched the show—or that they even *liked* it (for many did not)—our intention here is to highlight *El Chavo del 8*'s continued resonance. Today, the show exists in a liminal space—both here and not here—like a specter. This "always-in-between" (Deleuze, 1995) space, however, is not *a*temporal. In the Bergsonian sense, the past is never just the past; it always *coexists with itself as present* (Deleuze, 2004, p. 29, original emphasis). Following this logic, we propose that the show continues to exert extraordinary power and force over past and contemporary (re)formulations of *El Chavo* and the bodies entangled within its assemblages.

Coda

The conversations and connections sparked by this book are far from over. As such, a sense of "*unfinishedness*" (Waterhouse, 2012, p. 181, original emphasis) is to be expected. This feeling is not a sign of failure, but precisely how we wanted this rhizome book to work. For those unfamiliar with *El Chavo del 8*, we hope this text provides you with an entry point into the show, and for dedicated viewers—whether you were a devotee, skeptic, or apathetic—we hope we challenged any assumptions or fixed ways of thinking about *El Chavo del 8*.

As for providing recommendations for future areas of work or research, we are wary of this (customary) temptation for it risks stifling creativity and experimentation. Instead, we posed multiple questions along the way and leave you with a couple more: in a show that personally impacted each of this book's authors, how is *desire*[7] working? Or, to be more specific, what is the "desiring machine" (Deleuze and Guattari, 1987) of *El Chavo del 8*? How does desire move through this rhizome book, the show itself, its viewers, and the assemblages connected to/from/within? Relationships of desire, power, interest, and *El Chavo del 8* offer budding paths.

And, because we must end, we leave you with a Nietzschean adage that Deleuze often reiterated: a great thinker is someone who picks up an arrow launched by an earlier individual and shoots it in a different direction (Rajchman, 2016). With the help of our mediators, we hope this rhizome book has provided you with multiple arrows. We invite you to pick up an arrow—or perhaps a combination of them—and launch them anew.

Notes

1 David Graeber (2015) persuasively articulates how the bureaucratization of universities in the United States has "strangled" all "imagination" and "creativity" (p. 134). Graeber (2015) states: "There was a time when academia was society's refuge for the eccentric, brilliant, and impractical. No longer. It is now the domain of professional self-marketers. As for the eccentric, brilliant, and impractical: it would seem society has no place for them at all" (pp. 134–135). Furthermore—and particularly relevant for this book given how much we draw from his work—Graeber (2015) argues that if contemporary incarnations of Gilles Deleuze, for example, were to appear in the U.S.

academy today, he would likely never make it through graduate school. Even if a contemporary Deleuze *did* find a way to miraculously finish his degree and land a job in academia, he would probably never achieve tenure (Graeber, 2015).

2 This is the costume worn by one of Roberto Gómez Bolaños's characters, *El Chapulín Colorado.*

3 According to Rajchman (2000), Deleuze's philosophy is about making connections. These connections are not just to philosophy itself, but also to the human and nonhuman, organic and inorganic, material and immaterial realms.

4 As a reminder to the reader, on October 2, 1968, between 30 and 300 students and civilians were killed by Mexican military and police in the Tlatelolco district of Mexico City.

5 For Deleuze and Guattari (1986) Kafka was an initiator of "minor literature." They credit him with giving literature a "new face," changing "both its addresser and its addressee" (Bensmaïa, 1986, p. xiv).

6 Deleuze does not mask his contempt for television shows or clichés.

7 We refer to desire not in the Lacanian sense of lack or dearth, but in the Deleuzoguattarian logic, as a productive, moving force.

References

Bensmaïa, R. (1986). "Foreword: The Kafka Effect." In G. Deleuze and F. Guattari (Eds). *Kafka: Toward a minor literature.* Minneapolis: University of Minnesota Press.

Deleuze, G. (1989). *Cinema 2: The time image.* London: Bloomsbury Publishing.

Deleuze, G. (1995). *Negotiations.* New York: Columbia University Press.

Deleuze, G. (1997). *Essays critical and clinical.* Minneapolis: University of Minnesota Press.

Deleuze, G. (2006). *Two regimes of madness: Texts and interviews 1975–1995.* New York: Semiotext(e).

Deleuze, G. and Guattari, F. (1986). *Kafka: Toward a minor literature.* Minneapolis: University of Minnesota Press.

Deleuze, G. and Guattari, F. (1987). *A thousand plateaus: Capitalism and schizophrenia.* Minneapolis: University of Minnesota Press.

Deleuze, G. and Guattari, F. (1994). *What is philosophy?* New York: Columbia University Press.

Deleuze, G. and Parnet, C. (2007). *Dialogues II.* New York: Columbia University Press.

Foucault, M. and Deleuze, G. (1977). "Intellectuals and Power." In D.F. Bouchard (Ed.). *Language, counter-memory, practice: Selected essays and interviews.* Ithaca: Cornell University Press.

Graeber, D. (2015). *The utopia of rules: On technology, stupidity, and the secret joys of bureaucracy.* New York: Melville House.

Massumi, B. (1992). *A user's guide to capitalism and schizophrenia: Deviations from Deleuze and Guattari*. Cambridge, MA: MIT Press.

Popkewitz, T. S. (2005). "Introduction." In T. S. Popkewitz (Ed.). *Inventing the modern self and John Dewey: Modernities and the traveling of pragmatism in education* (1st edn pp. 3–36). New York: Palgrave Macmillan.

Rajchman, J. (2000). *The Deleuze connections*. Cambridge, MA: MIT Press.

Rajchman, J. (2016). "Deleuze's Nietzsche." *Nietzsche 13:13* [Web log]. Available at: http://blogs.law.columbia.edu/nietzsche1313/john-rajchman-deleuzes-nietzsche/.

Tomlinson, H. and Burchell, G. (1994). "Translators' Introduction." In G. Deleuze and F. Guattari. *What is philosophy?* (pp. vii–1). New York: Columbia University Press.

Wallin, J. J. (2012). "Bon Mots for Bad Thoughts." *Discourse: Studies in the Cultural Politics of Education, 33*(1), 147–162.

Waterhouse, M. (2012). "Book Review: A Deleuzian Approach to Curriculum." *Journal of the Canadian Association for Curriculum Studies, 10*(2), 174–182.

Glossary

In this addendum we list some of *El Chavo*'s most popular catchphrases, explaining their usage and meaning, in order to support readers unfamiliar with the show. Many of the phrases include words that were invented by Roberto Gómez Bolaños (most of which started with "ch" in his trademark style), and thus have no literal translation. While there are dozens more catchphrases, we focus on those used in the book.

- *¡Bueno, pero no te enojes!*: "Ok, but don't get mad!" or "Calm down!" was a strategy used by many in the cast to de-escalate situations.
- *¡Chanfle!*: Perhaps the most difficult word to explain, this made-up term was used to express surprise or puzzlement, and featured in Chespirito's skits.
- *¡Es que no me tienen paciencia!*: "The problem is that you have no patience for me!," always stated with frustration after someone yelled at El Chavo for not understanding something.
- *¡Fue sin querer queriendo!*: The closest translation would be "It was an accident!," usually uttered after an unintentional physical mishap.
- *¡Tenía que ser el Chavo del ocho!*: "Of course it was El Chavo!" was a sentence used to accuse El Chavo, after it was perceived—rightly or wrongly—that he was at fault for some mishap.
- *Eso, eso, eso, eso*: Always accompanied by the repetitive pinching of the thumb and index finger, this expression meant "That's it!" or "That's what I meant!"
- *Chusma or ¡No te juntes con esta chusma!*: While the word means "rabble," the longer phrase "Don't get close to the rabble!" was often used by Doña Florinda as a warning to her son, Quico.
- *Se me chispoteó*: Usually uttered by El Chavo, this made-up sentence always came after he blurted out something inappropriate or out of turn and meant "I didn't mean to say that."

Index

abandonment, of children 45, 46
adjustment, to crisis ordinariness 122–4
adulthood, and childhood 41, 45
adults
  acting as children 42, 52
  and children 59
  in the classroom 25–6
  violence of 44
affective realm, of children 46–7
Affective Turn 12, 113, 115
agency, and change 123–4
Aguasaco, Carlos vii, 11 (*see also* Plateau II - 4)
Ahmed, Sara 121–2
albur 23
Alharthi, A. 89
alienation 5
alterity
  childhood as 60, 62
  education as 63
Althusser, L. 137
Americanization, of popular culture 5, 22
Appadurai, A. 53, 54
Apple, M. W. 23
Arata, Nicolás vii, 10 (*see also* Plateau I - 1)
Argentina, *conventillos* 9
Ariès, P. 38, 41

Bauman, Zygmut 8
behaviors, deviant 43, 46
Berger et al. 5
Berlant, Lauren 113, 114–15, 119
Beverly Hills 90210 29, 30
Bhabha, H. 52, 55, 57, 59, 60, 62
border crossings, between identities and experience 104–5
borders, cross-border migration 5
Brazil
  and *El Chavo del 8* 51–2
  television 64n.2
Bruner, J. S. 94

Cabrera, Dulce María vii, 13 (*see also* Plateau III - 8)
Calderón, Felipe 151
Cannella, G. S. 42
Cantinflas 39
Cantinflas, Mario Moreno 162–3
capital, cultural/economic 3–4
capitalism, and education 21
Capps, L. 94
Caraballo, Limarys vii–viii (*see also* Plateau II - 5)
Caracazo, El 111–12
Carbajal Romero, José viii, 13, 157 (*see also* Plateau III - 8)
Carol Burnett Show, The 70
Carvalho, Ana Paula Marques de viii, 10 (*see also* Plateau I - 3)
catchphrases, used in *El Chavo del 8* 136, 187
censorship, Cuba 100–1, 107n.9
Cervantes Saavedra, Miguel de 70
change
  and agency 123–4
  and learning 29
Chaparrón Bonaparte 3
Chapulín Colorado 136
characters (*see also* names of individual characters)
  contradictions of the 55
  of Gómez Bolaños 3, 35, 136–7
  production of 61
chavo, term 71
Chespirito. *See* Gómez Bolaños, Roberto
Chespirito 3
child abuse 44, 46–7, 76–7, 79, 80, 83
child labor 78–9, 83
childhood(s)
  acting out 55–60
  and adulthood 41, 45
  construction of 10, 38, 48, 53
  discourses of 45–6, 49
  and education 60–4

and *El Chavo del 8* 75–82
imagining from adulthood 38–40
Latin American 19
meaning of contemporary 55, 56–7, 58, 61
neglect of 46–7
as Other 60
politics of 41
production of 36, 40
as a protected stage of life 44
social construction of 10, 38
and social inequalities 49
social norms of 42
and subalternity 63
in the *vecindad* 41–8
Western 41
children
abandonment of 45, 46
and adults 59
affective realm of 45–6
child abuse 44, 46–7, 76–7, 79, 80, 83
child labor 78–9, 83
circulating images of 53–5, 56
cultural production of 43
identity of 55
neglect of 44, 45
as Other 38, 41, 59, 63
rights of 44, 49
social discrimination of 62
Chile, and the United States 70
Chilindrina
bad behavior of 43
character of 3, 37, 49, 112, 117, 159
in the classroom 24, 25–6, 28
and Quico 166
Chimoltrufia, La 136
Chin, Elizabeth 3
citizen–state interfaces 145–6
citizenship, transcultural 89, 105
Clandinin et al. 104
class inequalities, and schools 21 (*see also* inequalities)
class tensions 120–1
classed identities 102
classrooms
adults in 25–6
represented in *El Chavo del 8* 31
as a stage 23
collective action 82
Colmenares, Erica viii, 12 (*see also* Coda; Plateau II - 6)
colonial presence 58
comedy
intergenerational appeal of 98, 103
Mexican 91
in Pepito 92–3
use of 39–40
comic climates 79–80
communication
and meaning making 52
and printed materials 69–70
connections, to *El Chavo del 8* 177–9
conventillos 9
Creation of the Media: Political Origins of Modern Communications, The 69
crises
in education 144
political crises in Mexico 6
state crisis 144
crisis ordinariness 114–15, 116, 117–19, 122–4
cross-border migration 5
cruel optimism 119–22
Cuba
censorship 100–1, 107n.9
and *El Chavo del 8*/*El Chapulín Colorado* 96
el solar 101, 102
and Pepito 92–3, 100, 102–3, 104, 105
and the United States 70
cultural capital 3
cultural icons, El Chavo/Pepito as 12, 88, 90
cultural production, of children 43
culture(s)
Anglo-global pop culture 32
cultural identity 104
and curricula 61
documentary 71
El Chavo del 8 in cultural flows 53–5
of Mexico 22
popular culture. *See* popular culture
of poverty 102
in transition 89, 94
urban 168
curricula
curricular politics 60–4
El Chavo del 8 as rhizomatic curriculum 7–9
Cvetkovich, Ann 123–4

de Certeau, M. 158–9, 168
De panzazo 151
Deleuze, G. 7, 9, 179, 180, 181, 182
depoliticizing effect, of Gómez Bolaños's TV 146
Derrida, Jacques 56, 58, 135, 138, 145
development discourse 5
deviant behavior 43, 46
Dewey, John 9
difference, production of 57
discourses
  breaking with 59
  of childhood 45–6, 49
  development discourse 5
  of innocence 42
  of Latin American brotherhood 5
  popular discourse 23
discrimination, social 62
discursive decline, and political decline 145
dissemination of meaning 135, 137
documentary culture 71
Don Quixote 70, 72
Don Ramón
  abusive behavior of 44, 80
  and Argentina 9
  character of 3, 42, 101, 114, 117, 122–3
  and Chilindrina 37
  and Doña Clotilde 120
  and Doña Florinda 26, 80, 102, 162, 166
  evasion of rent 117, 118
  revenge of 26–7
Doña Clotilde 4, 120, 167
Doña Florinda
  character of 3, 102, 117
  and Don Ramón 26, 80, 102, 162, 166
  and El Chavo 79
  and Profesor Jirafales 78, 120, 167
  and Quico 118

economic capital 4
economic liberalization policies, Latin America 113, 114–15
economics lesson 27–9
education
  as alterity 63
  and childhood 60–4
  crisis in 144
  media education. *See* media education
  and power 21
  and subalternity 63
  through the media 171
educational system 148
El Buscón 71, 74, 75–8, 80, 82–3
El Caracazo 111–12
El Chapulín Colorado 3, 150
El Chapulín Colorado (book) 71
El Chapulín Colorado (TV program)
  goodwill 96–9
  humor 103, 105
  innocence 96–9
  and Latin America 98
  nostalgia 87, 96–9, 103
  politics 99–101
  poverty 101–3
  social class 101–3
  word play 99–101
El Chavo
  accused of theft 97–8
  appeal of 166
  character of 3, 4, 22, 37, 88, 117
  components of 160–1
  as a cultural icon 88, 90
  deviant behavior of 46
  expulsion from the *vecindad* 46–7
  name of 74
  personal story of 72
  repetitions used by 136
  schooling of 22, 23–9
  staging of childhood 56, 58
  violence towards 80
  who is he? 159–60
  as young worker 79
El Chavo Animado 152
El Chavo del 8
  criticisms of 48
  IMDb entry for 91
  importance of 35, 155
  viewing figures 173n.7
  what was it? 3–6
El Chavo del Ocho. *See El Chavo del 8*
El Chómpiras 3
El diario de El Chavo del Ocho 72
El Doctor Chapatín 3
El Lazarillo de Tormes 69, 70, 71, 73, 75, 78, 79, 82
El Periquillo Sarniento 71, 73, 80, 82, 83
equivalence 138–9
exceptionalism, Venezuela 115–16
expectations, in the *vecindad* 42–5

experimentation
and adjustment 122–4
in *El Chavo del 8* 183

Fanon, Frantz 62
fantasy 139–40
Fernández de Lizardi, José Joaquín 73
fetish, images as 57
figured worlds 88, 89
filiation, contested norms of 45
14 y Medio 92, 106n.6
Frangella, Rita de Cássia Prazeres viii, 10 (*see also* Plateau I - 3)
Friedrich, Daniel viii, 10 (*see also* Coda; Plateau I - 1)

gender, and *El Chavo del 8* 43
Generación Y 92, 106n.6
girls, cultural production of 43
Giroux, H. A. 20
Glee 29, 30
global cultural flows 54
Global Initiative to End All Corporal Punishment of Children 80
globalization 8, 54, 89
Godínez 27, 28, 29, 31
Gómez Bolaños, Roberto
across cultural and national borders 95–6
characters of 3, 35, 136–7
conflicted relationship with Mexico 181
creativity and humor of 90–1
dissemination, interpellation, and re-interpellation of his TV 134–9
and *El Chavo del 8* 1, 38, 72
emergence of his TV 133–4
fantasy in the TV shows 139–42
narrative inquiry of his shows 93–6
politics and his TV 150–2
production of the state in his TV 142–6
significance of his work 20, 32, 87
success of *El Chavo del 8* 31
and *Televisa* 5
TV shows of 70, 131–2
good life fantasy 113, 114, 119, 122, 123
goodwill, and *El Chavo del 8*/*El Chapulín Colorado* 96–9
Guattari, F. 7, 9, 179, 181, 182
Guerra, J. C. 89, 104–5
Gullestad, M. 39

Hall, Stuart 20
history class 23–5
history, production of 32
Hockey, J. 41
Holland et al. 88
humor
and *El Chavo del 8*/*El Chapulín Colorado* 103, 105
as a medium 39
in transcultural contexts 88–90
hunger 77–8, 98, 101, 162 (*see also* poverty)

I Love Lucy 70
identification(s)
as an analytical scheme 137
cultural frames of 170
with the El Chavo character 159
El Chavo del 8 as a site of 117–19, 155
popular identification 12
identity(ies)
of children 55
and citizenship 105
classed 102
cultural identity 104
of El Chavo 28
identity formation 137–8, 141–2
as iteration 62
Latin American 70, 89, 95
production of 60
transcultural 88–9, 104, 105
Illich, Ivan 21
images
as fetish 57
as open text/double text 58
indigenous foreigner 8–9
industrialization 5, 163
inequalities
and childhood 49
comedians portraying 40
and schools 21
innocence
discourse of 42
and *El Chavo del 8*/*El Chapulín Colorado* 96–9
institutions, criticisms of 81, 83, 143
intergenerational appeal, of the comedy 98, 103
intermezzo, and media education 157–9, 171
International Education Center, Cuernavaca 21

Internet Movie Database (IMDb) site 90–1
interpellation 137–8
intimate public, *El Chavo del 8's* 12, 113–14, 118, 119, 124

Jackson Albarrán, Elena 19
Jaimito the Mailman 81
James, A. 41

Kafka, Franz 182
Kincheloe, J. L. 35, 45
knowledge, cultivated/official 23

La Bruja del 71. *See* Doña Clotilde
La vida inútil de Pito Pérez 71, 73–4, 82
Lacan, J. 137
Laclau, E. 137, 138
language
  albur 23
  popular language 162–3
  and reality 7
Latin America (*see also* names of individual countries)
  child labor in 78–9, 83
  childhoods in 19
  citizenship in 89
  economic liberalization policies in 113, 114–15
  and *El Chapulín Colorado* 98
  and *El Chavo del 8* 8–9, 48, 98
  identity 70, 89, 95
  politics of 4–5
  popular culture 104
  schooling in 31–2
  television in 70, 83n.4
  and United States 70
  and violent behavior 80
Latin American brotherhood, discourses of 5
learning, and change 29
Leichter, H. J. 95
lifestyle, the marginal as 170–1
"Los Caquitos" 136
"Los ricos también loran" 141
love, the search for 119–20

Mafalda 22
Marxist social theory 7
mass media
  audiovisual 70
  against mass media 169–70
  and the production of society/state 144
  and social inclusion/exclusion 39
  subversion of 168
massacre of students, Plaza de las Tres Culturas de Tlatelolco 6, 22
Massumi, Brian 13, 123–4
Mast, Gerald 79
meaninglessness, threat of 5
meaning(s)
  of childhood 55, 56–7, 58, 61
  dissemination of 135, 137
  and El Chavo character 159
  meaning making and communication 52
  -scapes of 54–5
  shared meanings 137
  and signifiers 135
media, and Americanization 22
media education 155–9
  education against the media 168–9, 171
  education in the media 161–2, 171
  education through the media 171
  mass media against mass media 169–70
  provided by *El Chavo del 8* 166, 167
media literacy 158
media signifiers 132
mediascapes 54–5
messages, conveyed by the sketches 98
Mexicanos Primero 151
Mexico
  comedic performers 91
  culture of 22
  demographics of 133–4
  Mexicanity 22
  Mexico City 163–4
  National Action Party (PAN) 151, 152
  neoliberal policies 102
  Partido Revolucionario Institucional (PRI) 6, 99, 134, 150, 152
  political crises 6
  politics 134
  television 134 (*see also Televisa*)
migration, rural to urban 5, 163
Miller, J. 60
Milmo, Emilio Azcarraga 141
mimicry 57, 58
minor literature 182–3
mobility 8
modernization 5
Monsiváis, Carlos 32

Montes de Oca, Marco Antonio 23
Moreno, Mario 39
Mouffe, Ch. 138

names, changes of 74
narrative, and self 94
narrative inquiry, of Gómez Bolaños's shows/Pepito's jokes 93–6
National Action Party (PAN) 151, 152
National Union of Education Workers (SNTE) 144
neglect
  of childhood 46–7
  of children 44, 45
neoliberal policies, Mexico 102
Ñoño Barriga 3–4, 24, 27, 28, 37–8, 44, 48, 49, 75
nostalgia, and *El Chavo del 8*/*El Chapulín Colorado* 87, 96–9, 103

Ochs, E. 94
Other
  childhood as 60
  children as 38, 41, 59, 63
  representation of 56, 57

parenting, contested norms of 45
parents, in the classroom 25–6
Parra-Moreno, Victoria ix, 10 (*see also* Plateau I - 2)
Partido Revolucionario Institucional (PRI) 6, 99, 134, 150, 152
pelados 91
Peña Nieto, Enrique 151
Pepito
  across cultural and national borders 95–6
  appeal of 87–8
  and Cuba 92–3, 100, 102–3, 104, 105
  as a cultural icon 12, 88, 90
  humor/sarcasm in 92–3
  nostalgia relating to 98–9
  political agenda of 99–100
Pérez Jiménez, Marcos 115
picaresque novels 11, 69–70
  and *El Chavo del 8* 71–83
Pinar, W. 63
Plaza de las Tres Culturas de Tlatelolco, massacre of students 6, 22
political crises, Mexico 6
political decline, and discursive decline 145
politics
  of childhood 41
  curricular politics 60–4
  depoliticizing effect of TV 152
  and *El Chavo del 8*/*El Chapulín Colorado* 99–101
  and Gómez Bolaños's TV 142–3, 150–2
  and humor/sarcasm 92–3
  Latin American 4–5
  Mexico 134
  National Action Party (PAN) 151, 152
  Partido Revolucionario Institucional (PRI) 6, 99, 134, 150, 152
  and Pepito 102–3
  and *Televisa* 150–2
Popis 26–7
Popkewitz, Thomas 8–9
popular culture
  Americanization of 5, 22
  concept of 20
  and cultural identity 104
  and *El Chavo del 8* 32, 161
  Latin America 104
  nature of 11–12
  and residual ideas 71
  suspect nature of 7
  and TV 133, 168
popular discourse 23
popular identification 12 (*see also* identification)
popularity, of *El Chavo del 8* 70, 124
Portelli, A. 94
poverty (*see also* hunger)
  comedians portraying 40
  culture of 101–2
  and *El Chavo del 8*/*El Chapulín Colorado* 101–3
  Mexico 134
  Venezuela 116
power
  and adult authority 41
  and curricula 62
  and education 21
  and popular culture 20
Profesor Jirafales
  character of 42, 83
  in the classroom 22, 26, 27, 80, 81, 147
  and Doña Florinda 78, 120, 167

economics lesson 27–9
frustrations of 148
history lessons of 8, 23–4
repetitions used by 136
progress
absence of in *El Chavo del 8* 30–1
and modern schooling 29
public institutions, limits of 82, 83
public pedagogy, and popular culture 20
Puerto Rico, and *El Chavo del 8*/*El Chapulín Colorado* 97
punishment, physical 76–7, 79, 80 (*see also* child abuse)

Quevedo, Francisco de 74
Quico
abusive behavior towards 44, 80
character of 3, 37, 49
and Chilindrina 166
and class issues 102
in the classroom 24, 25–6
and El Chavo 97–8
sending away of 118
Quiñones, N. 91

Rancière, J. 146
reality, and language 7
recognition
of children 62
and representability 57
representability, and recognition 57
reviews, of *El Chavo del 8* 91
Reyes, Dorian Flores 172
rhizomatic perspective, *El Chavo del 8* 6–9
rights
of children 44, 49
of young workers 79
role models 136
Romero, José Rubén 73
Ronzón, Ernesto Treviño ix, 12 (*see also* Plateau III - 7)
Rousseau, Jean-Jacques 41
rural to urban migration 5, 163

Sanchez, Yoani 92–3, 105, 106n.6
sarcasm 92
Saussure, Ferdinand de 138
-scapes 54–5
schism 57
schooling
as absurd 30
anti-progress class 29–31
economics lesson 27–9
of El Chavo 22, 23–9
and *El Chavo del 8* 77–82
history class 23–5
Latin American 31–2
school(s)
of *El Chavo del 8* 146–50
and inequalities 21
as simulacrum 30
sedimentation of meaning 135
self
making of 47
and narrative 94
self-construction 94
Señor Barriga 3, 80, 101, 117
signifiers, and meanings 135
Silva, A. 165
Simon, R. 20
sin querer queriendo 155, 156, 165
social, apolitical treatment of the 145
social class
and *El Chavo del 8* 44–5, 48
and *El Chavo del 8*/*El Chapulín Colorado* 101–3
social construction, of childhood 10, 38
social contexts, equivalence between 139
social continuity 146–50
social discrimination, of children 62
social groups, cultural/material difference of 48
social hierarchies, in the *vecindad* 26
social inclusion/exclusion, and the mass media 39
social inequalities, and childhood 49
social issues
comedians portraying 40
and *El Chavo del 8* 37–8, 47–8
social mobility 120
social norms, in the *vecindad* 42–5
social regulation, in the *vecindad* 147
social relationships, making of 47
social segregation, Venezuela 116
social services 82
social theory, Marxist 7
social welfare 114
socio-state dimension, Gómez Bolaños's TV 142–6

solar 101, 102
Sosenski, Susana 19
South Park 30
sovereignty, regional 5
Spivak, G. C. 63
Starr, Paul 69–70
state crisis 144
state, obturation of the 146–50
statehood, production of 143, 146
Stavans, I. 99, 104
Steinberg, S. 35
stereotypes
  of students 78
  and TV 141, 150
students, as teacher 29
subalternity, and childhood/education 63
subversion, of mass media 168
subversiveness, adults acting as children 52
Süssekind, M. L. 63
symbols, of representation 58

teachers, students as 29
telenovela, Venezuelan 118, 125n.1
Televisa 5–6, 70, 72, 91, 99, 131, 141, 142, 150–2
television 134–9
  Brazil 51–2, 64n.2
  dissemination, interpellation, and re-interpellation of Gómez Bolaños's 134–9
  emergence of Gómez Bolaños's TV 133–4
  fantasy in 140
  fantasy in Gómez Bolaños's TV shows 139–42
  in Latin America 70, 83n.4
  Mexico 134 (*see also Televisa*)
  politics in Gómez Bolaños's TV shows 142–3, 150–2
  and popular culture 133, 168
  production of the state in Gómez Bolaños's TV shows 142–6
  significance of 131
  and stereotypes 141, 150
  TV shows of Gómez Bolaños 70, 131–2
  Venezuela 113
The Simpsons 30, 73
Tlatelolco massacre 6, 22
transcultural citizenship 89, 105
transcultural identities 88–90, 104, 105
traveling library 8–9

United States, and Latin America 70
urban culture 168
urbanization 5

vecindades (housing projects)
  aesthetic experience of 167, 171, 173n.9
  childhoods in 41–8
  culture of poverty 102
  description of 4
  and *El Chavo del 8* 3
  life in the 164–5
  origins of 164
  in popular culture 20
  as a setting 135
  social hierarchies in 26
  social norms/expectations in 42–5
  social regulation in 147
  spatial arrangement of 9
  and Venezuelan viewers 8
Velarde, López 23
Venezuela
  El Caracazo 111–12
  El Chavo del 8 8, 112–13, 124
  exceptionalism 115–16
  poverty 116
  telenovela 118, 125n.1
  television 113 (*see also telenovela*)
viewing figures, *El Chavo del 8* 173n.7
violence (*see also* child abuse)
  adult 44
  in *El Chavo del 8* 79–80

Waiting for Superman 151
welfare systems 82
Williams, G. 39
Williams, Raymond 71
word play, and *El Chavo del 8*/*El Chapulín Colorado* 99–101

Žižek, S. 139–40

CPSIA information can be obtained
at www.ICGtesting.com
Printed in the USA
LVOW13*1625190318
570349LV00009B/180/P

9 781474 298902